THE SUN'S INFLUENCE ON CLIMATE

....

Princeton Primers in Climate

....

David Archer, *The Global Carbon Cycle*
Geoffrey K. Vallis, *Climate and the Oceans*
Shawn J. Marshall, *The Cryosphere*
David Randall, *Atmosphere, Clouds, and Climate*
David Schimel, *Climate and Ecosystems*
Michael Bender, *Paleoclimate*
Andrew P. Ingersoll, *Planetary Climates*
Joanna D. Haigh and Peter Cargill, *The Sun's Influence on Climate*

THE SUN'S INFLUENCE ON CLIMATE

Joanna D. Haigh and Peter Cargill

PRINCETON UNIVERSITY PRESS   *Princeton and Oxford*

Published by Princeton University Press,
41 William Street, Princeton, New Jersey 08540

In the United Kingdom: Princeton University Press,
6 Oxford Street, Woodstock, Oxfordshire OX20 1TW
press.princeton.edu
All Rights Reserved

Library of Congress Cataloging-in-Publication Data
Haigh, Joanna D.
The sun's influence on climate / Joanna D. Haigh and Peter Cargill.
pages cm. — (Princeton primers in climate)
Includes bibliographical references and index.
ISBN 978-0-691-15383-4 (hardback) — ISBN 0-691-15383-3 (hardcover) —
ISBN 978-0-691-15384-1 (paperback) 1. Solar-terrestrial physics. 2. Climatic
changes—Effect of solar activity on. 3. Weather—Effect of solar activity on.
I. Cargill, P. (Peter) II. Title.
QB539.T4H35 2015
551.5'271—dc23
2014038583

British Library Cataloging-in-Publication Data is available
This book has been composed in Minion Pro and Avenir
Printed on acid-free paper. ∞
Printed in the United States of America
10 9 8 7 6 5 4 3 2 1

Contents

THE SUN'S INFLUENCE ON CLIMATE

....

1 INTRODUCTION

A COMPREHENSION OF HOW THE SUN AFFECTS THE Earth is a fundamental requirement for understanding how climate has varied in the past and how it might change in the future. This is particularly important in the context of determining the cause(s) of climate change: we need to understand natural factors to be able to attribute to human activity any past or potential future influence on a range of timescales.

The extent to which the Sun drives changes in the Earth's climate has been the subject of speculation, scientific investigation, and, often, controversy over many centuries. Solar energy maintains the equitable global temperature, while the distribution of insolation across the globe results in night and day, and geographic variations in weather and climate as well as their seasonal modulation. The fundamental role of the Sun in climate is therefore undeniable. The question that arises is whether and how solar activity varies over time and how possible variations might be affecting our environment.

Naked-eye observations of sunspots have been made since ancient times. Babylonian and Chinese astronomers in the seventh century BC recorded dark spots on the face of the Sun, and court astrologers in ancient China believed sunspots foretold important events. In

Greece in the fourth century BC Theophrastus, in his book on weather forecasting, suggested that black marks on the Sun predicted rain. Intermittent sunspot sightings were recorded during subsequent centuries, but it was not until the telescope was invented around 1600 that routine observations were made. The common interpretation then was that these spots were planets transiting the face of the Sun, but it was Galileo who noted that the varying shape and speed of each spot belied that possibility and indicated that they were in fact blemishes on the solar surface. Galileo made a number of sketches from his observations of sunspots, an example of which is shown in Figure 1.1 alongside a recent image of the Sun. Galileo's picture shows the dark centers of the spots (the umbra) and the lighter surrounding regions (penumbra).

Later observers, including William Herschel in London at the end of the seventeenth century, noted that the number of sunspots was not constant but varied between none and many, and it was he who carried out what was probably the first scientific study of the relationship between sunspot number and weather. His publication of 1801 identified five periods of a few years each in the interval 1650–1717 during which sunspot numbers (as compiled by French astronomers) were low. Herschel then examined records of the price of wheat during that span and argued that high prices corresponded to product scarcity, which must have reflected poor growing conditions. He acknowledged that this was a somewhat indirect measure of temperature but reasoned,

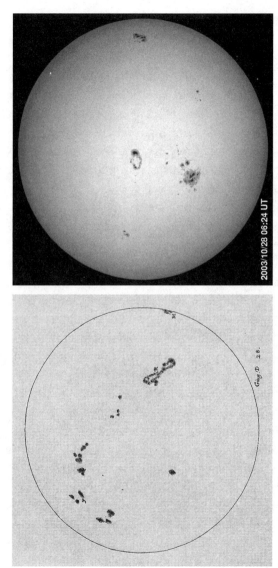

Figure 1.1 Left: An example from June 28, 1613, of Galileo's drawings of sunspots. (Used with the kind permission of Owen Gingerich.) Right: An image of the Sun in visible light acquired on October 28, 2003, by the MDI instrument on the SOHO spacecraft. (From http://sohowww.nascom.nasa.gov/)

pragmatically, "for want of proper thermometric observations, no other method is left for our choice." He concluded, "It seems probable that some temporary scarcity or defect of vegetation has generally taken place, when the sun has been without those appearances which we surmise to be symptoms of a copious emission of light and heat," that is, poor growing conditions when the Sun was less active. The statistical robustness of Herschel's work does not bear much scrutiny, but it set the scene for much that followed.

In the mid-nineteenth century the existence of the *sunspot cycle*, a semiperiodic waxing and waning in the observed number of sunspots, was discovered by a German apothecary and amateur astronomer, Heinrich Schwabe. Inspired by this finding, Rudolph Wolf, a Swiss schoolteacher, started to collate sunspot data and designed a system for intercalibrating observations, which is now called the *Wolf number*. He showed that the period of Schwabe's cycle varied between 8 and 17 years but averaged 11.1 years; records of the Wolf sunspot number now extend from the seventeenth century until the present day (see, e.g., Fig. 1.2). At about the same time as Wolf was doing his work on the sunspot cycle, several scientists noted that the Earth's magnetic field—measurements of which had been initiated by Carl Friedrich Gauss in 1835 and carried out at a number of observatories—varied almost in tandem. Studies had already been done on the relationship between geomagnetic storms and observations of auroras (northern lights), and the discovery of the sunspot cycle led to a

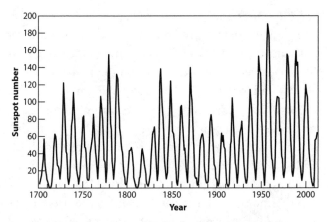

Figure 1.2 Yearly sunspot number for the years 1700–2013. (Data from WDC- SILSO, Royal Observatory of Belgium, Brussels, http://sidc.oma.be/silso/)

search for periodicities in auroral observations. It was understood that auroras were high in the atmosphere, but it was thought that they presaged fine weather near the Earth's surface, so the logical next step was to look for relationships between solar activity and weather.

A large number of studies ensued, whose major focus was on surface temperature, but other meteorological parameters of interest were precipitation (including evaporation, floods, droughts, river flows, and lake levels); atmospheric pressure; cloud cover; the position of storm tracks; the intensity and frequency of tropical storms and monsoons; temperature and winds at different levels in the atmosphere; as well as indirect measures such as the frequency of forest fires and shipping losses.

Notable among the studies were papers by Charles Meldrum, a Scottish meteorologist and government observer in Mauritius, who analyzed data from 144 stations across the globe over seven solar cycles in the mid-nineteenth century and found a solar cycle signal in rainfall; he also claimed "a strict relation" between sunspot number and the frequency of cyclones in the Indian Ocean. Another Scot, Charles Piazzi Smyth, Astronomer Royal for Scotland, counted meteorology among his many interests. He analyzed data acquired in the period 1837–1869 from four thermometers buried at different depths (to remove transient/diurnal signals) into the rock of Calton Hill in Edinburgh and has been widely quoted as finding significant correlation of temperature with sunspot number. That was a correct observation, but, interestingly, Smyth also noted "several numerical circumstances which show that the sunspots cannot be the actual cause of the observed waves of terrestrial temperature," demonstrating the skepticism which is unfortunately lacking in many works on apparent solar–climate links.

During that time the Indian subcontinent became part of the British Empire, and the government felt a concomitant responsibility to nurture agricultural productivity in that region. The severe impacts of Indian famines led to a focus on meteorology involving studies by, among others, Norman Lockyer (solar physicist, discoverer of helium, and founder of the journal *Nature*). Initially having traveled to India to observe a solar eclipse, Lockyer became interested in solar–climate links and compiled a list of all correlations established by researchers between sunspots, geomagnetism, temperature, and rainfall, focusing on cyclones and their effect on shipping.

Despite the large number of reports of Sun–weather links, by the end of 1870s there was a rise in criticism of the work. For example, a leading Indian government meteorologist stated in an official report to the famine commissioners that he could find no simple correlation. Concerns raised were based largely on some rather poor statistical relationships but also on the lack of any robust physical mechanisms that could account for the supposed association. The most plausible explanation would be a relationship between sunspot activity and solar energetic output. William Herschel believed that higher numbers of sunspots were associated with greater solar emissions of heat and light, but there was no evidence to support this hypothesis. Indeed, some believed that a preponderance of dark areas on the solar surface would result in reduced energy output.

Measurements of the Sun's radiant energy became the life work of Charles Greeley Abbott, director of the Smithsonian Observatory and secretary of the Smithsonian Institution in Washington, DC, during much of the first half of the twentieth century. He held a firm belief that solar irradiance varied and thus influenced weather, despite the claim by many climatologists that total solar irradiance (TSI) or "the solar constant" was just that. He set up carefully calibrated radiometers on mountains across the United States and in other countries, including Argentina, Chile, Egypt, and South Africa, and made measurements over nearly 40 years. At the same time many other scientists around the world were making similar measurements, and they produced a wide range of estimates of TSI: values published between 1900 and 1950 ranged from 1322 to 1465 W m^{-2}. No one was able to establish a repeatable link between sunspot number and irradiance, and instruments on balloons and rockets did not prove helpful. Finally, the launch of radiometers on Earth-orbiting satellites in the late 1970s made it possible to remove the effects of the intervening atmosphere and to show that TSI does vary, by a small fraction, in phase with sunspot number.

Despite some decline in enthusiasm for establishing solar–climate links at the end of the nineteenth century, work continued on a number of fronts, and papers continued to be published on statistical correlations. Over the next half century a confusing picture emerged with claims of in-phase and out-of-phase relationships, some of which switched behavior between periods. In 1957 the American Meteorological Society expressed its frustration with the statement that "[s]ome few results have been obtained that are highly suggestive. But none of these studies have produced any conclusive evidence that relationships do indeed exist."

While changes in irradiance provide the most likely candidate for any solar effect on climate, the lack of reliable measurements prior to the satellite era means that "proxies" such as sunspot number or geomagnetic indexes are still needed as indicators of past solar activity. Useful proxies can be found in isotopes such as carbon-14 (^{14}C) and beryllium-10 (^{10}Be) in ocean sediments and other geologic strata, ice cores, and tree rings. These isotopes are produced by reactions with galactic cosmic rays, whose flux into the Earth's magnetosphere is modulated by the strength and extent of the solar magnetic field. Thus the concentration of cosmogenic nuclides is inversely related to solar activity and can provide records for hundreds of thousands of years, although dating becomes problematic for the oldest samples.

The Schwabe cycle had been the focus of many studies, but longer-term variations in solar activity were also noted. Around 1887 the German astronomer Gustav Spörer identified a period circa 1645–1715 during which very few sunspots were observed, and this finding was confirmed in England shortly afterward by Edward Maunder, after whom the period was named the Maunder Minimum. Around eighty years later this topic was reinvestigated by U.S. solar physicist Jack Eddy, who used ^{14}C measurements to extend the solar activity record back to the eleventh century, identifying a Spörer Minimum in activity around 1500, another grand minimum around 1350, and a Medieval Maximum during the twelfth and thirteenth centuries. Eddy further claimed that the ^{14}C record provided an (inverse) measure of solar irradiance; thus low isotope levels would indicate a warmer Earth, and vice versa. He noted that the coldest excursions of the Little Ice Age coincided approximately with the Maunder and Spörer Minima, and the Medieval Climate Optimum with the solar Medieval Maximum.

Significant effort in contemporary climate science is invested in the attribution of causes to the general increase in global average temperature over the past one and a half centuries. It is very difficult to make a robust scientific case for anything other than a rise in the concentration of anthropogenic greenhouse gases as the major source of the warming; however, it is important to establish the roles of other factors—both those resulting from human activity (such as industrial particulates) and, importantly, those for which humanity cannot be held responsible (such as solar activity and volcanic eruptions). For this reason the evidence for Eddy's relationship has been the subject of much further investigation. Consideration of the Little Ice Age and Medieval Climate Optimum as global phenomena, rather than signals pertaining to northwestern Europe, has been challenged, and one interesting avenue of study is the influence of the Sun on regional, rather than global, climate.

The availability of high-performance computers, advances in climate modeling, and construction of comprehensive and robust records of meteorological variables are all helping in the quest to identify the extent to which the climate responds to changes in the Sun and to understand the physical mechanisms underlying these signals.

In this book we present some of the background to these endeavors. We start with an overview of the Earth's climate system—its composition, structure, and circulation—and some of the ways in which these vary naturally with time. We then look at key features of the structure of the Sun, its magnetic field, and atmosphere, and its emission of radiation and particles. In the next chapter we focus on solar radiation and its interaction with the terrestrial atmosphere in the context of the Earth's radiation budget and radiative forcing of climate, as well as its direct impact on atmospheric composition and temperature. We follow this discussion with a review of the temporal variation of several measures of solar activity. In the next section of the book we cover the evidence for an influence of solar variability on the atmosphere, oceans, and climate on timescales ranging from minutes (space weather) to thousands of years. We also consider the processes that might be responsible for the observed changes.

The subject of this volume is wide-ranging and sometimes controversial, and given space limitations, we cannot hope to provide a comprehensive coverage. Nevertheless, we hope the reader will take away a flavor of the science behind this complex and fascinating topic and of the challenges remaining to be addressed.

2   THE EARTH'S CLIMATE SYSTEM

·····················

THE EARTH'S ATMOSPHERE IS A THIN LAYER OF GAS surrounding the planet. Its composition, temperature, and motion are determined by internal chemical and physical processes as well as by complex interactions with other parts of the climate system—notably the oceans, cryosphere and biosphere—as illustrated in Figure 2.1. The energy source driving the climate is radiation from the Sun, which is balanced, on a global average, by the emission of heat (infrared radiation) to space. The Earth's radiation budget and details concerning the absorption of solar radiation are discussed in Chapter 4. In this chapter we outline the main features of the structure and circulation of the atmosphere to provide a background for discussion of the solar-induced anomalies in Chapters 6 and 7.

COMPOSITION

Below about 100 km in altitude the composition of the atmosphere is quite uniform, because turbulent mixing occurs on shorter timescales than does molecular diffusion, which would allow sorting of lighter from heavier molecules. The deviation of gaseous concentrations from

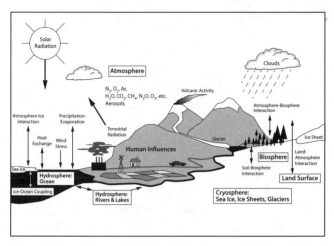

Figure 2.1 Components of the Earth's climate system. (Adapted from Cook, 2013, fig. 1.1) [1]

uniformity is due to the presence of sources and sinks. At the Earth's surface these can be the result of human activity, but within the atmosphere they are due to physical processes (such as photodissociation) or chemical reactions. Above 100 km the atmosphere is so rarefied (here the pressure is about one-millionth its value near the surface) that vertical mixing is controlled by molecular diffusion. One result of the low density is that at these very high altitudes the Sun's ionizing radiation produces long-lived free electrons. These create a charged layer called the *ionosphere*, which plays a key role in determining the Earth's electric field. In this book, however, we are concerned mainly with the climate of the lower atmosphere.

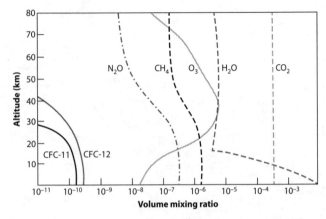

Figure 2.2 Concentration profiles of radiatively active species. The volume mixing ratio (vmr) indicates the number of air molecules of the specified gas as a fraction of the total number of air molecules. (Salby, 1996)[2]

In the lower atmosphere the composition is dominated by nitrogen (N_2, constituting 78% of the number of air molecules), but it plays a negligible role in interacting with solar (or infrared) radiation and is not involved in any chemical processes. The next most abundant constituent is oxygen (O_2, 21%), which, although chemically inert, is photodissociated by solar ultraviolet (UV) radiation into highly reactive oxygen atoms and thus plays a key role in determining the chemical composition of the stratosphere, and in particular of ozone (O_3). The vertical concentration profiles of other radiatively active gases are shown in Figure 2.2.

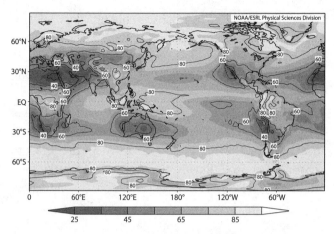

Figure 2.3 Percentage cloud cover in January (1871–2012). (Data from http://www.esrl.noaa.gov)

CLOUDS

A key component of the atmosphere is water vapor. Near the surface the water vapor concentration, which varies widely, is around a few percent but drops sharply with height (see Fig. 2.2) owing to declining temperatures. The colder temperatures cause the water to condense out and form clouds of liquid water droplets or ice crystals. Clouds play a crucial role in the radiation, heat, and water vapor budgets of the planet.

Figure 2.3 shows the distribution of cloud cover in January. Most cloud cover occurs in regions where air is ascending. When air rises it expands, owing to the drop in pressure, resulting in adiabatic cooling. The cooler air has a lower saturation vapor pressure and, at a given humidity, becomes saturated and water droplets condense out. The two major factors causing ascent are convection and convergence; an important example of the latter occurs in the intertropical convergence zone (ITCZ; see Fig. 2.5), where the northeast and southeast trade winds meet near the equator. The ITCZ moves north and south across the equator during the year, following the Sun, and is associated with intense storms and precipitation. Another example of forced ascent occurs in midlatitude (around 50°) weather systems that form when polar and subtropical air masses collide, forcing the warmer air to ride over the colder, creating weather fronts. These can be identified as the storm tracks in both hemispheres in Figure 2.3.

Convection occurs when air overlying the surface is heated by the ground, which is warmed by the Sun, or when air flows over a warmer surface. The warmed air becomes less dense than the surrounding air and starts to rise and cool. At some height (called the *condensation level*) the air becomes saturated, and cloud forms. Within the cloud the decrease of temperature with height is generally less than in clear air, so the ascending volumes remain warmer and more buoyant than the surrounding air. Given enough energy these convective clouds can develop into deep storm clouds, reaching the tropopause, where the wind blows ice "anvils" off the tops. The most intense convective clouds occur over hot continental land surfaces, as in equatorial Africa and South America. Clouds are also prevalent in the Asian monsoon, which occurs over the Indian subcontinent and parts of southern China during the summer months.

Cooling the air to saturation is not, however, a sufficient condition for cloud to form. Indeed, it is possible for relative humidities to reach values up to 500% without any condensation taking place, because water vapor requires a suitable surface, called a *condensation nucleus*, on which to condense. Normally, the condensation nucleus is not a liquid water surface (because the thermodynamics of that situation requires prohibitively high relative humidities) but, rather, particulate matter suspended in the atmosphere (called *aerosols*), such as sea salt, mineral dust particles, sulfates, or soot. The concentration and composition of atmospheric aerosols vary geographically. For example, sulfate aerosol is more abundant in the Northern Hemisphere, as it is generated in industrial regions. For the same total water content, a region with a high concentration of condensation nuclei will produce a larger number of smaller cloud droplets than a remote area with clean air, which will produce fewer, larger droplets. Smaller drops are more effective at scattering radiation (see chapter 4), so the cloud produced in the air with more aerosols has a higher albedo. It has been suggested that ionization of atmospheric aerosols by galactic cosmic rays may make them more efficient cloud condensation nuclei, thus potentially providing a route whereby solar modulation of cosmic rays could influence global albedo (see Chapter 7).

TEMPERATURE

At the Earth's surface, temperatures are generally highest over land surface in the tropics and in the summer subtropics, and lowest near the winter pole. The presence of cloud moderates the temperatures so as generally to keep surfaces warmer in winter and cooler in summer. Within the atmosphere the temperature varies with height, and different layers are traditionally named according to their vertical temperature gradients. Figure 2.4 identifies these layers in a global mean plot of the temperature of the atmosphere below 100 km.

The lowest portion is called the *troposphere*; here the structure is determined largely by adiabatic and convective processes, with the evaporation and condensation of water vapor playing a key role. The resulting temperature profile decreases with height up to the tropopause at 10–17 km altitude. Above this lies the *stratosphere*, where temperature increases with altitude to the stratopause near 50 km. Here water vapor and condensation are not important; heating is largely due to the absorption of solar UV radiation by O_2 and O_3, which is balanced by infrared (IR) radiative cooling. Above the stratopause the UV heating declines, along with temperature, into the mesosphere.

It is common in meteorology to measure altitude not only by geometric height but also by atmospheric pressure, as indicated by the two vertical axes in Figure 2.4. The altitude of a particular pressure level depends on the temperature of the atmosphere below it (warm air occupies greater volume), so that the relationship between geometric height

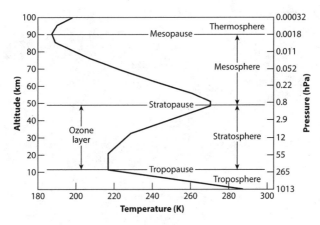

Figure 2.4 Global annual mean atmospheric temperature as a function of height (km) and air pressure (hPa). (Houghton, 1977)[3]

and pressure varies, and the values given in the figure are only representative. In practice the altitude of a given pressure level, called its *geopotential height*, is often used as a measure of the mean temperature below it. Note further that the pressure at any point indicates the weight of the atmosphere above it, so it can be seen from the figure that the troposphere, which contains almost all the cloud and all the weather and climate as experienced by humankind, contains about 80% of the entire atmospheric mass.

WINDS

Figure 2.5 shows the mean wind fields for January at 200 hPa (near the tropopause) and near the surface. The surface trade winds can be seen as easterly/equatorward

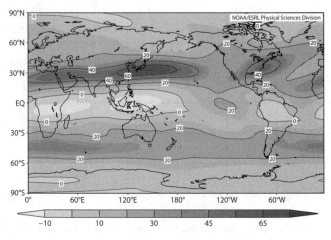

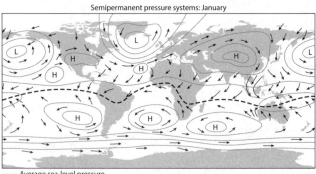

Semipermanent pressure systems: January

——— Average sea-level pressure
——→ Surface wind-flow patterns

Figure 2.5 Top: Zonal wind (m s^{-1}) for January at 200 hPa (~12 km altitude). (From http://www.esrl.noaa.gov). Bottom: January mean surface wind (arrows) and surface pressure (contours). Also shown is the position of the intertropical convergence zone (black dashed line). (Laing and Evans, 2011)[4]

flow in the tropics, converging south of the equator at this time of year into the ITCZ. In midlatitudes the surface flow is predominantly westerly/poleward, although this flow is disrupted by the presence of land masses in the Northern Hemisphere. At 200 hPa the key features are the midlatitude jet streams, with the strong flow east of North America and Asia marking the North Atlantic and North Pacific storm tracks, respectively.

The wind and temperature structures of the atmosphere are closely related through what is known as the *thermal wind relationship*: for example, where temperature decreases toward the pole, westerly wind increases with height above that level, and vice versa. This relationship, which relies on a balance between the horizontal pressure gradient and the force arising from the Earth's rotation, exists throughout the atmosphere away from the surface, where frictional drag becomes significant, and the equator, where the rotational effects vanish. Thus in the lower troposphere the difference in temperature between the warm tropics and the cool poles is associated with the westerly jets, which peak near the tropopause. Tropospheric convective processes cool the tropical tropopause region, so that at this level the temperature is a minimum near the equator and maximum at the summer pole, and the westerly jets decline in strength. In the upper stratosphere, above about 30 hPa, the temperature increases uniformly from the winter to the summer pole, and the thermal wind balance results in a strong westerly jet in the winter hemisphere and an easterly jet in the summer hemisphere.

As a consequence of the thermal wind relationship any changes in the Sun that affect atmospheric temperature will necessarily influence the wind structure as well.

TRANSPORT OF HEAT BY CIRCULATION OF THE ATMOSPHERE

On a global and annual average the solar energy absorbed by the Earth is balanced by thermal infrared radiation emitted to space. However, solar radiation absorption has a strong latitudinal variation, with higher values near the equator and lower values near the poles, while the outgoing infrared radiation has only a weak latitudinal dependence. Thus there is a net surplus of radiative energy at low latitudes and a deficit at high latitudes. This differential heating tends to set up thermal convection cells with air rising near the equator (in the ITCZ); flowing poleward near the tropopause; then cooling, sinking, and flowing equatorward again near the surface. These overturning circulations, named *Hadley cells*, extend to about 30° latitude in each hemisphere; the latitudinal extent is determined by the point at which the flow becomes dynamically unstable owing to westerly accelerations resulting from planetary rotation.

The instability produces waves in the zonal (east-west) flow which are manifest in surface weather patterns— the low- and high-pressure systems familiar in forecast charts—and which provide the means for poleward heat transport in midlatitudes throughout the depth of the troposphere. Because of the steeper temperature gradients during winter, the intensity of these systems, and the amount of energy they transport, is greatest at that time of year.

The horizontal wavelength (around a line of latitude) of the waves is typically of order 1000–2000 km, so the waves are sometimes referred to as *synoptic-scale eddies*. Other instabilities in the atmosphere result in waves of much longer or much shorter wavelength. In particular, *Rossby waves*, which depend on the rotation and geometry of the Earth, are several thousand kilometers long and propagate upward, through regions of westerly winds, well above the tropopause, where they are responsible for driving the mean overturning circulation of the stratosphere. Here the temperature structure is associated with a mean meridional flow in which air entering the tropical lower stratosphere is transported toward both poles in the lower stratosphere but higher up forms a simple single cell that ascends near the summer pole and descends near the winter pole.

SURFACE PRESSURE

Surface winds are also closely related to surface pressure: winds blow with low pressure to the left in the Northern Hemisphere, and to the right in the Southern. Thus air circulates counterclockwise around regions of low pressure in the Northern Hemisphere and clockwise around high-pressure regions, with the opposite circulation in the Southern Hemisphere. (The term *cyclone* is used for low-pressure systems and *anticyclone* for high-pressure systems, in both hemispheres.) The wind–surface pressure relationship can also be seen on the global scale. Figure 2.5 (lower panel) presents the key features of sea-level pressure across the globe: low pressure exists near the equator and in the region of 60° latitude in both hemispheres, with high-pressure bands near 30°. Comparison with the surface winds shows this pattern is consistent with the low-latitude trade winds and midlatitude westerlies. In the Northern Hemisphere the Siberian High and Aleutian Low, associated with the distribution of the continents and mountainous regions, are persistent features of the Northern Hemisphere winter and responsible for steering weather systems to the north and south, respectively.

Areas of surface convergence and ascent (the ITCZ, the ascending branch of Hadley cells, and midlatitude storm tracks) are associated with low surface pressure and high precipitation, while descent and surface divergence are accompanied by high surface pressure and clear skies.

Changes in climate, whether introduced by human activity such as the emission of greenhouse gases, or natural factors like volcanism or solar activity, may involve significant local variations in wind or precipitation by a slight shifting in the positions of some of these key climate features, even if their impact on global mean temperature is very small.

OCEANS

The oceans also carry heat poleward, although outside the deep tropics their contribution is smaller than that due to the atmosphere. Ocean surface currents are wind driven, so there is a clear signature of the trade winds on either side of the equator. In each of the major ocean basins the horizontal flow is predominantly in an anticyclonic gyre with poleward flow on the west side of the basin and equatorward flow on the east. The western boundary currents bring relatively warm water to midlatitudes, thus ameliorating the climate in these regions. The wind stress also drives an upwelling of water in coastal regions that results in shallow (a few hundred meters deep) overturning circulations in the tropical ocean basins, varying in strength on timescales of decades. On much longer timescales the deep ocean circulation, which is driven by gradients of heat and salinity, involves flow throughout the global oceans. Surface water in the Pacific is heated by the sun and flows eastward across the Pacific and Indian Oceans then around the Cape of Good Hope into the Atlantic. It crosses the equator and travels northward into the North Atlantic, where much of its heat is lost to the atmosphere. The cold water then sinks and returns to the Pacific at depth via the Antarctic Ocean. Some studies have indicated that climate change might alter this "conveyor belt," with significant potential impact, especially on lands bordering the North Atlantic.

MODES OF VARIABILITY

The atmosphere exhibits a number of natural variations, or modes of variability, that can be important in determining local climate. Each mode has a characteristic geographic influence which fluctuates in strength on its own timescale. Note that this fluctuation does not imply two distinct climate states with the atmosphere switching between them but, rather, a distribution of states between the extremes. Climate forcing factors, including solar variability, may affect the frequency of certain phases of the modes. We do not attempt to review all such modes but to outline those that have been suggested to be associated with solar variability.

The El Niño–Southern Oscillation

The El Niño–Southern Oscillation (ENSO) phenomenon is the leading mode in the tropical Pacific, although its influence is felt globally. El Niño was the name given to the periodic warming of the ocean waters near the coast of Peru and Ecuador which adversely affects the fishing industry in this region by suppressing upwelling nutrient-rich deep waters. The Southern Oscillation was the term used to describe a periodic variation observed in the east-west gradient of surface atmospheric pressure across the equatorial Pacific. Associated with this pressure gradient is the *Walker cell*, an overturning circulation with eastward flow near the tropopause over the tropical Pacific and westward flow near the surface. It is now known that El Niño and the Southern Oscillation are parts of the same complex ocean–atmosphere interaction, referred to as ENSO. A typical configuration of sea surface temperature (SST) and sea-level pressure is shown in Figure 2.6 for the El Niño event of 1997. Positive anomalies of sea surface temperature initially appear

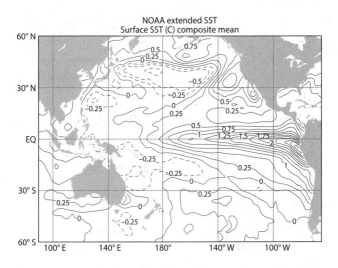

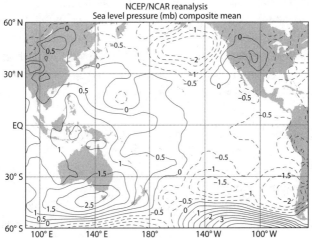

Figure 2.6 Anomalies in (top) sea surface temperatures and (bottom) surface pressure in the Pacific Ocean associated with the strong El Niño–Southern Oscillation signal in 1997. (From http://www.cdc.noaa.gov/ncep_reanalysis/)

on the east side of the equatorial Pacific and over a period of a few months spread westward until they cover most of the ocean at low latitudes, associated with a weaker Walker cell, producing the pattern shown in Figure 2.6. Concurrently, the region of maximum rainfall, normally over the maritime continent, shifts eastward into the Pacific. In the opposite phase (a so-called La Niña) a cooler-than-average SST anomaly appears in the east, associated with a stronger Walker cell.

ENSO events occur, irregularly, approximately every 2–5 years, and a wide range of climate anomalies in the extratropics as well as near the equator have been associated with this phenomenon. Some theories have provided insight into the complex coupling of the atmospheric and oceanic dynamics and thermodynamics which drive ENSO, but the explanation is not yet complete. There is some evidence (see Chapter 6) that high solar activity is associated with a La Niña–like response in tropical sea surface temperatures.

Polar Modes

At high latitudes the leading winter modes in the Northern Hemisphere are the North Atlantic Oscillation (NAO) and the Pacific–North America Oscillation (PNA). These are sometimes viewed as part of the same phenomenon referred to as the Arctic Oscillation (AO), which is associated with variations throughout the lower atmosphere in the Northern Annular Mode (NAM). In the Southern Hemisphere an Antarctic Oscillation

(AAO) at the surface is associated with the Southern Annular Mode (SAM). The NAM and SAM account for approximately one quarter of the variability in winds at high latitudes.

Over the Atlantic in winter the average sea-level pressure near 25°–45° N is higher than that around 50°–70° N (Fig. 2.5, bottom). This pressure gradient is associated with the storm tracks which cross the ocean and determine, to a large extent, the weather and climate of western Europe. It has long been known that variations in the pressure difference are indicative of a large-scale pattern of surface pressure and temperature anomalies from eastern North America to Europe. If the pressure difference is enhanced (Fig. 2.7, top), then stronger-than-average westerly winds occur across the Atlantic; and cold winters are experienced over the northwest Atlantic and warm winters over Europe, Siberia, and East Asia, with wetter-than-average conditions in Scandinavia and drier conditions in the Mediterranean. The fluctuation of this pattern is referred to as the NAO, and the pressure difference between, say, Portugal and Iceland can be used as an index of its strength. The AO is a zonally symmetric mode of variability characterized by a barometric seesaw between the north polar region and the midlatitudes in both the Atlantic and Pacific.

Values for the NAO index have been reconstructed dating to the seventeenth century from individual station observational records of pressure, temperature, and precipitation. The index shows large interannual variability but until recently has fluctuated between positive

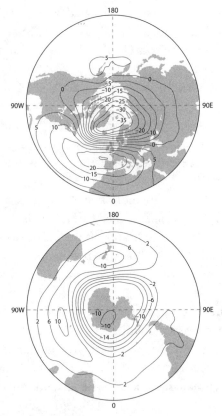

Figure 2.7 Top: Changes in winter (December–March) surface pressure (in units of 0.1 hPa) corresponding to a unit deviation of the NAO index over the period 1900–2005. Stronger west-to-east winds are associated with the steep pressure gradient in the eastern Atlantic. Bottom: The SAM 850 hPa geopotential height (m) (i.e., the temperature of the lowest ~1.5 km of the atmosphere) pattern regressed onto the SAM index. A positive SAM is associated with colder-than-average polar temperatures and warmer-than-average midlatitudes, producing a stronger polar vortex. (Redrawn based on IPCC AR4, figs. 3.30 and 3.32)[5]

and negative phases with a period of approximately four or five years. Throughout the 1970s and 1980s the NAO exhibited a positive increasing trend in its westerly phase, and it was suggested that this might be a response to global warming; however, that trend has subsequently reversed. Computer models of the general circulation of the atmosphere (GCMs; sometimes also called *global climate models*) are quite successful in simulating NAO-type variability but not the timings of particular phases. Some GCMs show increasing values of the NAO index with increased concentrations of greenhouse gases, whereas other studies suggest that the NAO pattern itself may be modified in a changing climate, so that use of simple indexes may not be appropriate.

In the Southern Hemisphere various definitions of SAM have been proposed; a commonly used one is the normalized difference in the zonal mean sea-level pressure between 40° S and 65° S. The sea-level pressure pattern associated with a positive SAM is nearly annular, with low pressure centered on the South Pole, and a ring of high-pressure anomalies at midlatitudes. This pattern is associated with stronger circumpolar westerlies and colder polar temperatures (Fig. 2.7, bottom).

An area of ongoing research is the extent to which coupling with the state of the stratosphere is important in producing realistic NAO/NAM and SAM patterns. The planetary-scale waves, which propagate upward in winter high latitudes, deposit momentum and heat that feed into the general circulation. Where the wave absorption takes place depends on the ambient temperature and wind structure, and any changes introduced into the mean temperature structure of the stratosphere may result in a feedback effect on lower atmosphere climate. This is a plausible mechanism for the production of polar mode signals in tropospheric climate by factors— specifically solar variability—which affect the heat balance of the stratosphere. Data and modeling studies have already shown such a response to heating in the lower stratosphere from volcanic eruptions, and there is some evidence (see Chapter 7) that low solar activity is associated with a negative NAO-like response.

The Quasi-Biennial Oscillation

The quasi-biennial oscillation (QBO) modulates winds and temperatures in the equatorial stratosphere. A given phase (east or west wind anomaly) starts in the upper stratosphere and moves downward toward the tropopause at a rate of about 1 km per month, to be replaced by winds of the opposite phase. The largest amplitude in the zonal wind variation occurs at about 27 km altitude; the cycle repeats about every 28 months. The QBO comes about because of interactions between vertically propagating waves and the mean flow. When the wind blows from the west (QBO west phase) westward-moving waves can propagate freely, but eastward-moving waves are absorbed and deposit their momentum, thus strengthening the existing westerlies and moving the westerly peak downward. Somewhat above this absorption layer the westward-moving waves are dissipated and weaken the west wind, eventually changing its direction to easterly. The absorption of the westward waves then starts to propagate downward, reversing the phase.

The effects of the QBO are not restricted to equatorial regions. The transports of heat, momentum, and ozone to high latitudes are all modulated. On average the westerly phase of the QBO is associated with colder winter temperatures at the North Pole in the lower stratosphere, because the ability of midlatitude planetary waves to propagate into the equatorial westerlies is enhanced, leaving the cold winter pole undisturbed. Some studies have suggested that this relationship is much more robust when the Sun is at lower levels of activity. Understanding this behavior, together with its coupling to the atmosphere below, may contribute to a fuller description of solar–climate links.

CLIMATE AND WEATHER

The climate of any geographic region is defined by the statistics of its behavior over a long (several decades) time period and is typically expressed as averages and standard deviations of the local meteorological conditions. Weather, in contrast, represents essentially instantaneous conditions, which are, of course, far more variable. To understand climate we do not need to know the detail of every individual weather event, but we do need to understand the bounds of normal variability and in assessing the significance of a climate change signal, take these into account on all timescales. Some modeling studies have suggested that global warming will be accompanied by more extreme events, so that extremely hot (or cold or wet or dry) periods may become more frequent. Such behavior needs to be assessed carefully in attempts to extract often small solar signals from "noisy" climate records.

CLIMATE MODELING

GCMs have become important tools for atmospheric and climate science. They are used by national meteorological agencies to produce daily weather forecasts and by climate scientists to understand the past and to try to predict the future.

Such models are computer programs which apply mathematical equations describing the basic laws of physics—conservation of energy and mass, Newton's laws of motion, thermodynamics of gases and liquids—to an atmosphere and oceans represented by continuous fluids resting on the surface of a rotating planet of prescribed shape. Solving these equations simultaneously on a 3D spatial grid produces fields of the key meteorological variables—including temperature, pressure, wind speed, cloud cover—which are then stepped in time into the future. For weather forecasting the models are repeatedly reinitialized with current measurements (from ground-based weather stations and instruments on satellites, airplanes, and ships); they generally do not include simulation of ocean circulations, because these change on longer timescales. For climate studies the models are initialized with the conditions pertaining at some point in time and then run forward with prescribed changes in parameters such as greenhouse gas concentrations or solar irradiance; over longer periods the coupling between atmosphere and oceans becomes important.

Because of inherent nonlinearities in the climate system the trajectory predicted by a GCM for its future state is sensitive to details of the prescribed initial conditions. Thus model predictions, for both weather and climate, are usually carried out as ensembles of many runs, each starting with a slightly different state. The results provide a measure of spread of likely outcomes around a mean trajectory.

Constraints on the Performance of GCMs

GCMs have been successful in representing the state and variability of weather and climate, but practical considerations limit their accuracy and capability. A major issue concerns the discretization of equations in space and time necessary to solve them numerically. The choice of grid size and time step determines the smallest and fastest features that can be resolved by the model, so that certain processes are necessarily excluded. For example, much atmospheric turbulence takes place on scales smaller than the grid of a climate model but is important in cloud development and the transport of chemical species. The large-scale effects of such small-scale processes are included in GCMs by *parameterizations* which adopt approximations, or use empirical relationships between model variables, to represent them. Furthermore, some aspects of the underlying physical and chemical processes are so complex that parameterizations are unavoidable. These include radiative transfer (of solar and infrared radiation) and cloud formation, discussed further in the next section.

At present some of the world's most powerful (civilian) computers are used for running GCMs, but even with these, resources such as processing speed, memory, and storage place fundamental limitations on what is possible in terms of increasing grid resolution or reducing the need for parameterizations. The necessity to produce ensembles of runs (see the preceding section) places further demands on resources.

In addition it is always possible that some relevant physical or chemical process is missing from a model. These are often discovered by carefully comparing model results with available observational data. Nevertheless, the ability of the models to realistically simulate current and past climate suggests that no major factor is currently being ignored.

Radiative Processes in GCMs

Radiative processes are fundamental in determining the thermal structure of the atmosphere and in driving global circulations. The interactions of solar radiation with the atmosphere, and the redistribution of energy by infrared processes, are the subject of more detailed discussion in

Chapter 4; here we just note the importance of accurately representing radiative processes in GCMs.

Numerous different schemes are used in GCMs to calculate radiative heating and cooling rates as well as photochemical processes. Determining the requirements of a particular study involves consideration of which gases to include, the representation of cloud, the spectral resolution of the calculation, and the vertical resolution of the grid.

The spectral resolution is especially pertinent to the representation of solar variability. Many GCMs include only a rudimentary representation of the stratosphere and of the radiative processes within it, but variations in solar UV radiation primarily affect this region, and solar effects on climate may be manifested by this route, transmitted through dynamic coupling to the atmosphere below (see Chapter 7). If that is the case, then the ability of such GCMs to simulate solar effects will be diminished.

Water and Clouds in GCMs

GCMs need to include representation of the hydrologic cycle, including the transport of water vapor within the atmosphere, evaporation and condensation, and clouds and precipitation.

Approximations in the representation of cloud formation and cloud radiative properties remain major causes of uncertainty in current GCMs and are the main reason for the wide range of values for the climate sensitivity parameter, outlined in Chapter 4. As discussed previously, cloud formation depends on factors ranging from local topography, large-scale flow, and the temperature and humidity of air masses to the microphysical composition of particulates in the atmosphere. It is not feasible for GCMs to account fully for all these factors, so they include a variety of parameterizations for cloud prediction. On the largest spatial scales and annual timescales these parameterizations are moderately successful. For local cloud on timescales of hours, however, the predictions can be quite poor. Even disregarding the problems associated with cloud prediction, the models still must calculate the effects of cloud on the radiation fields. Again, gross approximations have to be made in terms of the necessary integrations over spectral and spatial variables.

Also in a state of comparative infancy is the representation in models of land and ice surfaces and their interaction with radiation. It should be noted that these surfaces, along with cloud cover, determine the amount of solar radiation that the Earth reflects to space, so they could be relevant to studies of the impact of solar variability on climate.

Models of Reduced Complexity

For many applications simpler, less computationally expensive models may be more appropriate than the full GCMs just described. Models which focus on one (or more) aspects of a problem and disregard other features which are deemed to be less important in the particular context reduce computational demands and allow for greater experimentation and, possibly, easier interpretation of results.

A common approach is to reduce the dimensionality of the model at the expense of a complete representation of dynamic processes. For example, two-dimensional (latitude–height) models, in which zonal mean quantities are considered, have been successfully used in studies of stratospheric chemistry. One-dimensional (height only) energy balance models, which consider only transport from vertical diffusion, can provide a useful first-order estimate of the global mean response to radiative forcing perturbations.

Alternatively, if the focus is on a process occurring in a particular region, it may be useful to restrict the spatial extent of the model, although care must then be taken with boundary conditions. Another approach is to simplify the treatment of certain processes to focus on others. For example, studies of coupling between the tropical lower stratosphere and tropospheric circulations, in response to solar forcing of the stratosphere, have been carried out using a model in which all radiative and latent heating processes were reduced to a simple relaxation to a reference equilibrium temperature. This model fully represented dynamic processes but avoided the necessity for detailed radiative calculations, thus allowing numerous experiments to be carried out.

In the next chapter we turn to the Sun, looking at its structure and the fundamental processes driving variability in the components of its emissions that might affect climate.

3 THE SUN

Four general factors contribute to the Sun's potential role in variations in the Earth's climate.

(i) The fusion processes in the solar core determine the solar luminosity and hence the "base" level of radiation impinging on the Earth.

(ii) The presence of the solar magnetic field, and dissipation of associated electric currents, leads to radiation at ultraviolet (UV), extreme ultraviolet (EUV), and X-ray wavelengths which can affect certain layers of the atmosphere.

(iii) The variability of the magnetic field over a 22-year cycle leads to significant changes in the radiative output at some wavelengths and also modulates the incoming flux of galactic cosmic rays (see Chapter 5).

(iv) The interplanetary manifestation of the outer solar atmosphere (the solar wind) interacts with the terrestrial magnetic field, leading to effects commonly called space weather. (see Chapter 8)

The Sun is a fairly typical star, of spectral class G, midway through its lifetime. In the overall classification of stars, the Sun is said to be a *cool star*, with an effective temperature of 5785 K. Having formed from a dust cloud roughly 4.5×10^9 years ago, it is now burning its

Table 3.1
Solar Properties

Radius of Sun (R_S)	6.96×10^8 m $= 109$ R_E ($R_E \sim 6.37 \times 10^6$ m)
Distance to Earth (1 astronomical unit; AU)	1.5×10^{11} m $= 215.5$ R_S
Solar mass (M_S)	1.99×10^{30} kg $= 3.33 \times 10^5$ M_E ($M_E = 5.97 \times 10^{24}$ kg)
Average density	1.4×10^3 kg m^{-3}
Luminosity	3.86×10^{26} W
Effective temperature	5785 K

fuel, converting hydrogen into helium, and is on what is known as the *main sequence* of stellar evolution. Eventually, this steady burning will cease, and the Sun will move rapidly off the main sequence, undergo catastrophic expansion as a planetary nebula—in the process shedding its outer envelope—and end its life as a white dwarf. Table 3.1 summarizes important properties of the Sun, with terrestrial quantities (subscripted E) included for comparison. Background material on most topics in this section can be found in the online journal *Living Reviews in Solar Physics* (www. http://solarphysics.livingreviews.org/).

A QUICK TOUR

As a guide to the rest of this chapter, we summarize the structure of the Sun. The interior and atmosphere have several different regions, as shown in Figure 3.1 (a). Panel

(a)

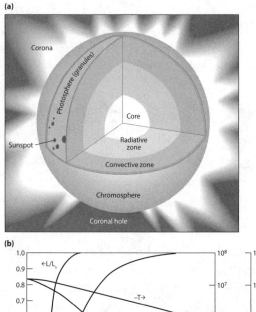

(b)

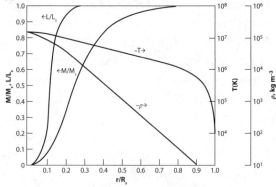

Figure 3.1 (a) A schematic of the solar atmosphere, showing the various regions discussed in this chapter. (NASA/Goddard Space Flight Center) (b) Structure of the solar interior. The two curves starting at the upper left show the temperature and mass density as a function of distance from the center of the Sun. The two curves rising from the lower left show the luminosity normalized with respect to that at the visible surface, and the normalized cumulative mass (i.e., $\int 4\pi r^2 \rho(r)\,dr/M_s$). (NASA/Marshall Space Flight Center).

(b) shows the properties of the ionized gas (plasma) in the solar interior.

The solar atmosphere comprises a number of distinct regions whose properties can be summarized as follows:

- The core (0–0.27 R_S): the location of nuclear fusion ($T \sim 15$ MK; density $\rho \sim 1.6 \times 10^5$ kg m^{-3} at the center)
- The radiative zone (0.27–0.7 R_S): region of outward energy transport by radiation (T decreases from 8 MK to 0.5 MK, and ρ decreases from 1.5×10^4 kg m^{-3} to 200 kg m^{-3})
- The convection zone (0.7–1 R_S): region of outward energy transport by convection (T decreases to 6600 K, and ρ decreases to of order 2.5×10^{-4} kg m^{-3} at the top)
- The photosphere: the location of temperature minimum (4300 K)
- The chromosphere (1–1.01 R_S): complex region above photosphere; hotter than photosphere (T up to 20,000 K; ρ falls to 10^{-8}–10^{-10} kg m^{-3})
- The corona (>1.01 R_S): tenuous hot outer atmosphere ($T > 1$ MK; $\rho \sim 10^{-11}$–10^{-12} kg m^{-3})
- The solar wind: expansion of the corona into interplanetary space

SOLAR RADIATION

It is useful to consider first the radiative output of the "Sun as a star," that is, as a single spatially and temporally unresolved entity. Figure 3.2 (a) shows the solar irradiance

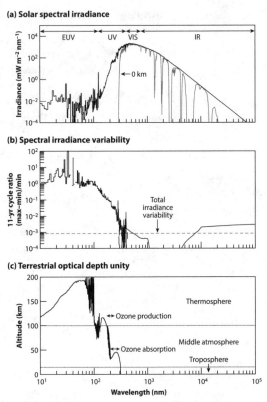

(a) Solar spectral irradiance

(b) Spectral irradiance variability

(c) Terrestrial optical depth unity

Figure 3.2 (a) Solar irradiance as a function of wavelength above the Earth's atmosphere (black curve) and at the surface (dashed curves). (b) The fractional difference between maximum and minimum of the solar cycle. The total variability is the horizontal dashed line. (c) A measure of the absorption of radiation in the atmosphere as a function of wavelength. The solid line indicates approximately the height above the surface that radiation of a given wavelength is absorbed. The atmosphere is split into three general regions, and the regions where ozone chemistry is important are highlighted. In all panels, wavelength is split into four broad domains: extreme ultraviolet (EUV), ultraviolet (UV), visible (VIS), and infrared (IR). (Lean and Rind, 1998)[6] as a function of wavelength. The solid lines represent the irradiance outside the Earth's atmosphere, and the dotted line, the irradiance at the surface. The units of the vertical axis are power per unit area per nanometer, and summing over all wavelengths gives the total solar radiation incident upon the Earth, the so-called solar constant or total solar irradiance (TSI). The best current estimate of TSI is 1361 W m^{-2}, a value somewhat lower than that, of order 1365 W m^{-2}, assumed in the recent past (see Chapter 5). The reader should be aware that some older figures in this book were prepared using the higher value.

The irradiance peaks at roughly 550 nm—not surprisingly, in the visible—and falls off rapidly in both the UV and infrared (IR). In the visible and near-IR the solid curve resembles blackbody radiation with an effective temperature of 5785 K. The IR is not discussed further. However, additional features not described by a blackbody become apparent in the UV and EUV owing to a range of atomic transitions in the atmosphere above the visible surface. Production of this radiation requires temperatures in the outer atmosphere regions (the chromosphere and corona) to be in excess of that observed at the visible surface. Indeed, the presence of the very hot solar corona was inferred in the late 1930s by observations of iron emission lines that can occur only at temperatures of order 1 MK.

The dotted line in Figure 3.2 (a) shows that the amount of radiation reaching the surface of the Earth falls off rapidly toward the UV. Radiation at these wavelengths is absorbed by higher regions of the atmosphere; Figure 3.2 (c) shows where this occurs, as defined by the location of an optical depth of unity. It is seen that most of the UV down to 100 nm is absorbed by the first 100 km of the atmosphere. A series of blocking wavelengths in the IR become more dominant in the far-IR. These are discussed further in Chapter 4. At longer wavelengths (radio), additional wavelength bands penetrate to the surface. Although these are important for observing the Sun and the Universe, the energy in this radiation is insignificant. (An important point concerning Figure 3.2 that we return to in Chapter 5 is that determination of the solar irradiance requires space-based observations, since significant wavelength ranges cannot reach the ground.)

Figure 3.2 (b) shows a further very important result, namely, the variability of the irradiance across a solar cycle, which is defined as the difference in spectral irradiance between the maximum and minimum of the cycle. The solar cycle is discussed more fully in Chapter 5, but we see here that while the total irradiance varies very little over 11 years (0.1%), there are large variations on the short-wavelength side, in the UV, EUV, and X-rays. At the shortest wavelengths, much of this variation is transient (e.g., owing to solar flares), but there is also a systematic increase in such radiation at the maximum of the solar cycle.

THE SOLAR INTERIOR AND VISIBLE SURFACE

The structure of the solar interior has been well understood in general terms for many decades and is summarized in Figure 3.1(b). (Shu provides a useful introduction to stellar structure.[7]) The figure shows the temperature and density as a function of solar radius. In general, the interior can be considered as a balance between a gravitational force directed inward and an outward pressure gradient, with the structure modulated by energy transport involving radiation and convection. Many generations of such solar models have been developed, with the interior structure constrained by the need to produce the observed luminosity.

As noted earlier, there are three regions in the interior. In the center is the core, where nuclear fusion takes place. Fusion occurs principally through proton-proton chains, with the carbon-nitrogen-oxygen (CNO) cycle contributing a very small amount to the energy release. The products of the fusion process are photons and neutrinos. Contemporary solar models put the extent of the core at 1.88×10^8 m, or 27% of the solar radius. By terrestrial standards the density of the core in particular is very large. Note in Figure 3.1(b) that the luminosity reaches its surface value before 0.3 R_S, evidence that in this particular model, fusion ends inside this distance.

Outside the core, the structure of the interior is determined by transport processes of the photons and hot plasma. Of importance is whether the atmosphere is *convectively stable*. Convective motions occur when a gas is heated from below and thus rises—a process familiar in many aspects of life. Immediately outside the core, the solar interior is stable to convection, and photons move outward very slowly by radiative transport through the *radiative zone*. Here the photons undergo continual

47

scattering by the hot, dense plasma, and their travel time through this region due to random-walk processes can be estimated in the range of tens of thousands of years.

The radiative zone extends as far out as 4.87×10^8 m, or 70% of the solar radius. The nature of the radiative zone, in particular the increase in the opacity of the plasma, is such that it has a high temperature gradient. In a stratified atmosphere such as the solar interior, there is a critical magnitude of the outward temperature gradient above which convection takes place, and the atmosphere is said to be *convectively unstable*. Why does this transition from radiative to convective energy transport matter? First, convection is a much faster way to move energy around. An element of plasma rising through the convection zone at 1 km s^{-1} will reach the solar surface in a few days, in contrast with tens of thousands years, the snail-like rate of energy transport in the radiative zone. Second, the convective flows, coupled with the differential rotation of the Sun described later, are believed to lead to the generation of the solar magnetic field.

The description of the solar interior to this point has been based on stellar structure models using solutions of hydrostatic atmospheres coupled to appropriate radiative physics. What is the evidence that they represent reality? The last three decades have seen a fundamental breakthrough in the study of the solar interior from an experimental perspective using the technique of helioseismology, which has enabled a vital verification of these models (e.g., Gizon et al., 2010[8]). The Sun, like any gas, can sustain oscillations which take the form of sound waves. The focus of research has largely been on what are called *p-* (pressure) *modes*. These are small pressure fluctuations, associated with sound waves, that lead to tiny oscillations of the solar surface (a few hundred kilometers in amplitude), with a broad spectrum of frequency in the range of a few megahertz. (A second class of predicted long-wavelength modes (g-modes) to probe the deep solar interior has proved more elusive.) Observations of p-modes, especially from space, lead to very tight constraints on the structure of the solar interior. In general, these observations have confirmed the three-part structure discussed earlier. In addition, through variations between the predicted and measured sound speed, helioseismology has identified a very interesting region at the base of the convection zone called the *tachocline* (by analogy with the oceanic thermocline), to which we will return shortly.

Another important result from helioseismology concerns the internal rotation of the Sun. It has long been realized that the solar surface rotates differentially: the equator rotates once every 25 days, and the polar regions take considerably longer—35 days. It has been accepted that the combination of convection and this differential solar rotation plays a key role in the generation of the solar magnetic field. Frequency splitting of p-modes due to rotation has enabled generation of maps of the rotation of the solar interior. It has been found that the radiative zone rotates roughly constantly, whereas the convection zone rotates differentially, with the rotation rate roughly constant along lines of constant elevation angle above and below the equator. The boundary between these regions, the tachocline, is thus the site of considerable shear in the internal rotation that is believed to be essential for magnetic field generation.

The visible surface of the Sun lies at the top of the convection region. It can be defined as the location where (visible) photons stop random walking and begin to fly, and is called the *photosphere*. A very small amount of this light can be absorbed by dense features higher in the atmosphere such as filaments (prominences), but almost all escapes unimpeded and travels into space. The temperature in the photosphere falls from about 6500 K to a minimum of 4300 K. Gas is rather weakly ionized at the photosphere, with <1% of the density in the plasma state.

The photosphere is very dynamic. We noted previously that the outer solar interior is in a state of convection. Consider for a moment a rising blob of gas. To conserve mass, after this blob rises to the surface, it must find a way back down, completing a convective cell. Videos of the solar surface show that it is very structured, with the required large-scale plasma flows occurring on a variety of scales. The most familiar of these are granules (1000 km in diameter and moving at 0.25 km s^{-1}) and supergranules (tens of thousands of kilometers; 0.3 km s^{-1}). Recent ground- and space-based measurements show these flows in remarkable detail. As we discuss next, an important aspect of these flows is that they push and concentrate magnetic fields, leading to the two most significant solar phenomena from the viewpoint of total solar irradiance (TSI), namely, sun-spots, and faculae.

THE SOLAR MAGNETIC FIELD

Before addressing sunspots and faculae, we need to discuss the solar magnetic field in more general terms. The first evidence for a strong solar magnetic field came from observations of sunspots by Hale in 1908, who identified kilogauss (0.1 T) magnetic fields from the Zeeman splitting of spectral lines. (Note that contemporary solar literature uses gauss almost exclusively as the unit of the magnetic field vector. However, to use SI units consistently throughout the text, we use tesla; $1 \text{ G} = 10^{-4} \text{ T}$.) To place this value in a terrestrial context, the Earth's surface field, although displaying variations in latitude, is of order $5 \times 10^{-5} \text{ T}$. The existence of a more general solar magnetic field was the subject of some debate in the distant past but is now universally recognized as present.

We noted earlier the existence of the tachocline, a region of strong velocity shear at the base of the convection zone. The evidence for the origin of the solar magnetic field at or near the tachocline is now strong on the basis of helioseismology, numerical modeling, and theory. However, what is far less certain is how a self-sustaining magnetic field that has the properties present in, for example, the sunspot cycle (see Chapter 5) can be maintained. This is referred to as the *solar dynamo problem*. The process whereby the generated field rises from the tachocline to the visible surface is better understood, having been first outlined by Parker.[9] An element of magnetic flux in a convecting atmosphere is buoyant, because the density inside the flux element must be smaller than outside, assuming equal temperatures owing to radiative effects. So such flux bundles can rise, and the rate of rise is determined by their field strength, the structure of the solar atmosphere, the strength of the convection, and the aerodynamic drag they experience.

The detection of surface magnetic fields relies on a spectroscopic technique known as the *Zeeman effect*. In the presence of a magnetic field, a spectral line is "split" (Zeeman splitting) from its field-free value, and the shifts in wavelength are well known. This technique works well in sunspots and is effective in the rest of the photosphere. In other parts of the solar atmosphere, however, Zeeman splitting is in competition with broadening of emission lines owing either to motions of the distribution of particles whose emission is being observed (thermal broadening) or to random (turbulent) motions of the medium itself (nonthermal broadening). It does not take much of an increase in temperature above photospheric values for the Zeeman effect to be swamped.

What does the solar magnetic field look like at the surface? Magnetic field maps of the photosphere (magnetograms) can readily be generated. Figure 3.3 (a) shows a typical example from the Helioseismic and Magnetic Imager (HMI) instrument on the NASA Solar Dynamics Observatory spacecraft. The image shows only the longitudinal field component, with black and white being inward and outward polarities, respectively. Such magnetograms are obtained from radiation from an Fe I line formed at photospheric temperatures (around 6000 K).

The two striking aspect of this image are the large structures (sunspots) discussed later and the large number of smaller regions of opposite polarities. Indeed, when the solar surface is examined with better spatial resolution, a hierarchy of structures becomes visible; there is a remarkably large range of magnetic structures, so that the subject of solar magnetic activity is by no means confined to sunspots. The distribution of magnetic flux in this range of structures follows a power law over many decades—a remarkable result.[10] Large sunspots with typical fields in excess of 0.1 T and dimensions of several million meters are the least common, while small flux elements, again with field strengths of order 0.1 T but scales of hundreds of kilometers, arise all over the solar surface. The flux distribution suggests a self-similarity in the physical process generating these magnetic elements and poses a stern test for dynamo theory and magnetic field transport in the convection zone. Also, the slope of the distribution suggests that all magnetic structures must be taken into account when considering the total solar magnetic field budget.

Finally, we note that since the magnetic field is a vector, measuring all three components is highly desirable. This is easier said than done and relies on the polarization signatures introduced by the Zeeman effect. The Stokes parameters are all measurable, in principle. The longitudinal field requires just two, whereas the transverse field (and hence the field vector) requires all four.

(a)

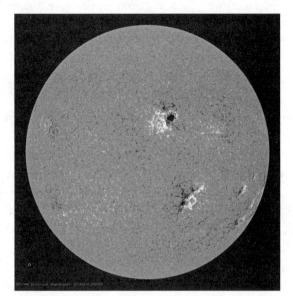

(b)

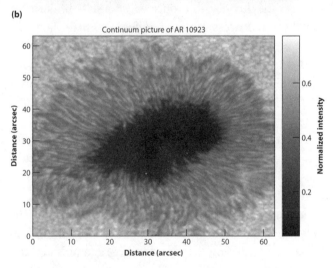

(c)

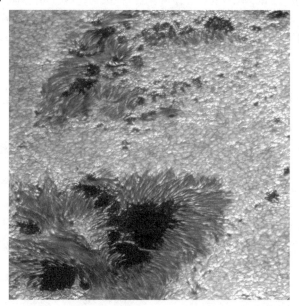

Figure 3.3 A composite picture of magnetic features on the solar surface. (a) A full-disk magnetogram of the Sun taken by the Helioseismic and Magnetic Imager (HMI) instrument on the NASA Solar Dynamics Observatory (SDO) spacecraft. Black and white are inward and outward polarities, respectively. The dominant structures are sunspot groups, with significant other structure visible elsewhere. (SDO/HMI) (b) A typical sunspot as seen by the optical telescope (SOT) on the Japanese Hinode spacecraft. The axes are in units of arcseconds (1 arcsec $\cong$ 725 km). (SOT/Hinode/JAXA) (c) Part of the solar surface as seen by the Swedish Solar Telescope (SST) with sunspots and faculae visible. The faculae are the numerous small, bright regions. The size of the image is approximately 53 × 53 Mm. (SST; Marshall Space Flight Center)

Modern instrumentation (e.g., the Advanced Stokes Polarimeter on the Japanese *Hinode* spacecraft and the HMI instrument on SDO) now makes vector field measurements feasible.

SUNSPOTS AND FACULAE

Sunspots have been well observed and documented for many centuries (e.g., Thomas and Weiss[11]). Figure 3.3 (b) shows a typical example. The dark region (the umbra) measures 20 arcsec (~15,000 km—big enough to swallow the Earth). It is surrounded by the penumbra (lighter gray on the image), which shows remarkable fine structure. Movies of such sunspots show these to be in constant motion, with flows (Evershed flow) and waves in continual evidence.

As shown in Figure 3.3 (a), sunspots are seen on the solar disk as dark regions. They are dark because they are cooler than the surrounding photosphere and so are not seen readily in visible light. Typical sunspot temperatures in the umbra are perhaps 2000 K lower than in the surrounding photosphere (so, roughly 3700 K). This lower temperature means that the solar surface in a sunspot is lower relative to the rest of the Sun (the *Wilson depression*), because in a hydrostatic atmosphere the density decreases exponentially. The characteristic scale is the *scale height*, a quantity proportional to the temperature. Thus cooler regions attain a surface density at lower heights than do warmer ones.

The formation and persistence of sunspots is related to the Sun's magnetic field. The presence of a strong vertical magnetic field in the umbra inhibits overturning convection (the field and plasma are frozen to each other); the plasma in the sunspot is thus not replenished by convection, and so it cools. In the brighter and hotter penumbra the magnetic field lies closer to the solar surface as the magnetic field structures itself to close in a neighboring sunspot of the group. Sunspot phenomenology such as lifetime, structure of sunspot groups, and motion is well documented. Contemporary modeling of sunspots does an excellent job of reproducing their basic structure.

Sunspots reduce the visible light from the Sun at 1 AU by a magnitude of order 2 W m^{-2} at the peak of the solar cycle to almost nothing at solar minimum. Many years ago, the question was posed, what happens to the outward solar radiation that is inhibited by the spots? This became known as the "bright ring problem" because the expectation was that the radiation would be deflected to appear around the spot. The very high efficiency of thermal energy transport in the convection zone resolved this problem, because it distributes the outward energy across the surface.

The second important photospheric structures from the viewpoint of TSI are *faculae* (Latin: little torch[12]). Figure 3.3 (c) shows part of the solar disk toward the west limb. There are a number of sunspots, but the new features to note here are the numerous small, bright regions. These are faculae and are associated with small regions of concentrated magnetic field. Because they are bright, they radiate more. In fact, the brightening due to faculae compensates for any darkening due to sunspots.

The role of the magnetic field is similar in some ways to that in sunspots. Pressure equilibrium across the boundary of these regions implies that the gas pressure must be lower in faculae than in the neighboring regions. In turn, this depresses the visible surface within the faculae. When viewed at an angle, as in this image away from disc center, the walls of these depressed regions allow radiation from the slightly hotter surrounding walls to escape from the Sun. As with sunspots, contemporary modeling can account for their basic structure.

CHROMOSPHERE, CORONA, AND SOLAR WIND

Figure 3.4 shows the temperature and density profiles above the visible surface. As expected in a stratified atmosphere, the density falls off rapidly. However, the temperature rises, first slowly through a region called the chromosphere (up to 20,000 K) and then abruptly into the corona (>1 MK). Why is this? Radiative heating can be ruled out, since the solar plasma becomes "optically thin" with height above the surface, so that solar radiation traveling outward cannot be reabsorbed and account for the heating. (This behavior differs from that in hot stars, where the surface radiation can be absorbed, driving massive stellar winds.) Note that the position of the boundaries between the different regions moves in response to the behavior of the entire system.[13]

The only alternative is that energy is being added directly to the chromosphere and corona, resulting in in

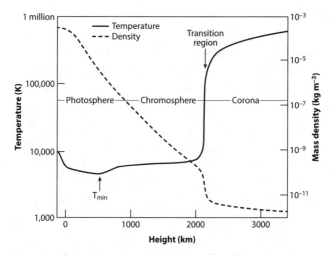

Figure 3.4 The temperature and density profiles of the solar photosphere and corona, showing the boundaries between the different regions. In reality the transition region does not occur at a fixed position but is defined by energy transport processes.

situ heating. Further, the only credible source of this energy is the magnetic field, because the magnetic energy density is much greater than anything else. However, in addition to being hot, the corona is an excellent electrical conductor, and dissipation of electric currents proves to be rather difficult unless very small scales are created. This is the focus of present-day research.

It was mentioned earlier that a fundamental difficulty exists with the measurement of magnetic fields above the visible surface, and this proves to be a major problem in understanding the corona. Techniques based on the study of coronal oscillations (*coronal seismology*) have provided estimates of order 2–4×10^{-3} T, which seems low. Measurements of radiation at radio wavelengths give alternative values of order 0.1 T but require subtle interpretation. Radiation in IR lines, combined with Stokes polarimetry, produces estimates in the outer corona of order 5×10^{-4}–10^{-3} T, but this information comes from radiation along an extensive line of sight. Images of the corona, such as from the Solar Dynamics Observatory, can provide information on the topology of the coronal field (Fig. 3.5), which arises because in a hot, highly conducting plasma such as the corona, plasma is said to be "tied" to the magnetic lines of force. The plasma can move freely along the magnetic field but not across it. This means that structures seen in, for example, SDO images are interpreted as outlining the coronal magnetic field.

The corona radiates over a wide range of wavelengths, ranging from hard X-rays during flares through soft X-rays and EUV in the nonflaring Sun. Because the radiation is emitted in a medium that is optically thin, photons escape from the corona without interacting with the plasma. Remarkable images can thus be obtained. Figure 3.5 shows a full-Sun image from NASA's SDO spacecraft. This is radiation from a transition of highly ionized iron (Fe IX), an emission line that gives strikingly clear images with the type of telescope flown on SDO.

Three regions can be identified. The bright parts, known as active regions, are regions where the magnetic field lines rise through, and close into, the photosphere. They are the sites of the most intense EUV and X-ray

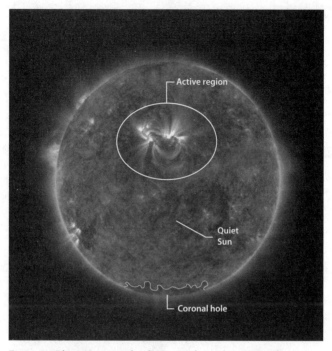

Figure 3.5 The Sun as seen by the Fe IX channel on SDO. This emission line is associated with plasma at around 1 MK. The three regions described in the text are outlined. (Cargill and De Moortel, 2011[14]; images courtesy AIA/SDO)

emission and are associated with significant sunspot groups. The dark regions are still hot (1 MK) but are dark because there is little plasma there. They are *coronal holes*, the origin of the solar wind (discussed later), and have a magnetic field that closes deep in interplanetary space. The rest of the image shows what is called the "quiet Sun," for obvious reasons.

Table 3.2
Properties of the Outer Solar Atmosphere

Parameter	Quiet Sun	Coronal hole	Active region
Coronal T (MK)	1.5	1	3
Coronal n (m^{-3})	5×10^{14}	2×10^{14}	5×10^{15}
Energy loss (corona; W m^{-2})	3×10^{2}	8×10^{2}	10^{4}
Energy loss (chromosphere; W m^{-2})	4×10^{3}	4×10^{3}	2×10^{4}
Mass loss (to solar wind; kg m^{-2} s^{-1})	Small	2×10^{-9}	Small

Source: Withbroe and Noyes (1977)[15]

What is the energetic significance of the chromosphere and corona in the context of the total solar luminosity? Some properties are summarized in Table 3.2. The chromospheric radiation exceeds that from the corona at all times, and one can calculate a typical combined flux at 1 AU by assuming that a small percentage of the Sun is covered by active regions and the rest is quiet Sun, which gives of order 10^{23} W, well under 0.1% of the luminosity. This is at the maximum of the activity cycle (see Chapter 5). So why do the chromosphere and corona matter? (a) As discussed in Chapter 7, there is now evidence that UV radiation plays a local role in determining weather patterns. (b) Short-wavelength radiation, especially when it is transient, can play an important role in the subject of space weather, creating times of abrupt ionization in the ionosphere. (c) The eruptive phenomena discussed shortly are a major aspect of space weather and the coupling of space with the terrestrial magnetosphere and ionosphere.

Other than via photons, the solar wind is the principal way in which the Sun is connected to the Earth and its environment. The solar wind is observed in interplanetary space as a very low density, hot (hundreds of thousands of kelvins), supersonic (hundreds of kilometers per second) magnetized plasma (a few nanotesla) that originates in regions of the solar corona whose magnetic field extends into interplanetary space. Why does the solar atmosphere choose to expand in this way? Suggestions of a medium between Sun and Earth that could transmit signals in a few days date to the nineteenth century (e.g., the Carrington flare), but the presence of a supersonic wind was predicted on largely theoretical grounds by Parker[16] and confirmed by in situ satellite measurements a few years later. The role of the solar wind in transmitting large disturbances is discussed shortly, but another important feature of the solar wind is embedded magnetic field turbulence. The variation of this through the solar cycle is discussed in Chapter 5, but we note here that the turbulence is "strong" in the sense that its amplitude is comparable to that of the background magnetic field, and the turbulence plays a significant role in scattering particles, including galactic cosmic rays.

FLARES, CORONAL MASS EJECTIONS, AND ENERGETIC PARTICLES

In addition to long-term variations in emissions from the outer atmosphere over a solar cycle (see Chapter 5), there are important short-term variations as well. The most important of these are solar flares[17] and coronal mass ejections.[18] This is not the place to rehash old arguments about which causes the other (if indeed there is any causality), but a satisfactory description for the largest flares is that the coronal magnetic field evolves to a state of disequilibrium and undergoes a violent disruption. Typical energy in these big flares is of order 10^{25} J. However, it is well known that flare energies follow a power-law distribution in energy (roughly $N(E) \sim E^{-1.5}$) over several decades, indicating some ubiquitous physical process. Smaller flares do not produce a major eruption, which suggests that the eruption is not essential to driving the main energy release, as is widely believed.

While flares are observed over a wide range of wavelengths, a fascinating and unsolved aspect of their physics is that a significant fraction of the flare energy release (up to 10s of percent) is initially in the form of energetic charged particles, particularly electrons with energies between 10 and 500 keV, though energized ions are also produced. The magnitude of the radiation (radio, hard X-ray, and gamma ray) from these particles is very small, although it still serves as an important diagnostic tool. However, the hot plasma produced by the thermalization of these particles radiates very efficiently in the X-ray and EUV. The lifetime of large flares is tens of minutes, and even longer-lived events lasting up to a day (such as seen in "post"-flare loops).

For large flares, a major part of the disruption is the ejection of plasma and magnetic field from the Sun into interplanetary space in the form of a coronal mass ejection (CME). CMEs are important for three reasons. First, the shock waves they produce are very effective accelerators of ions, which are a major space weather effect (see Chapter 8). Second, CMEs often have a strong, organized embedded magnetic field. As discussed further in Chapter 8, the interaction of the solar wind with the Earth is most effective when the interplanetary magnetic field (IMF) is southward, and CMEs can have long intervals (several hours) of quite strong (>10 nT) southward IMF. Third, the ejection of magnetic field into interplanetary space is an essential aspect of the change of polarity of the solar magnetic field. This observation was noted some time ago by B. C. Low, who argued that since the electrical conductivity in the solar atmosphere is so large, a viable way of changing the polarity is not to dissipate the field in situ but to fling it far away into interplanetary space.

4 SOLAR RADIATION AT THE EARTH

....

Solar radiation is the climate's fundamental energy source. In this chapter we consider the solar irradiance at the top of the Earth's atmosphere, its variation with location and season, and its energy distribution within the climate system. We discuss how changes in the radiation balance may influence global surface temperature and may thus be involved in climate change. And we investigate the radiative processes which influence the atmospheric temperature structure and some of the chemical processes, particularly those influenced by solar radiation, that determine atmospheric composition.

SOLAR RADIATION AT THE TOP OF THE ATMOSPHERE

To understand how variations in solar activity might affect climate it is necessary to look at the amount and distribution of solar energy reaching the Earth. If the Sun's output were constant, then the incident solar radiation would depend only on the distance between the two bodies. This distance varies during the year owing to the ellipticity of the Earth's orbit, measured by its *eccentricity* (e). For a circular orbit the value of e is zero; higher values

....

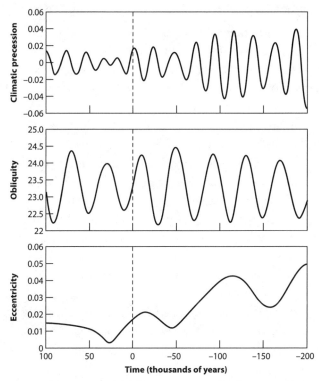

Figure 4.1 Time series of the three parameters that determine the Earth's orbit around the Sun: eccentricity, precession, and tilt. (Based on the calculations of Berger and Loutre, 1991)[19]

indicate a more elliptical orbit. The value of e varies in time, with periods of around 100,000 and 413,000 years, owing to the gravitational influence of the Moon and other planets (see Fig. 4.1). Its maximum value is about 0.058, while its current value is about 0.017. The ratio of the closest to the farthest distance of the Earth from the

Sun during a single orbit (year) is given by $(1 - e)/(1 + e)$, which at current values is about 0.967. Thus because the intensity of solar radiation falls off with the square of the distance from the Sun, currently about 7% less solar energy flux is received at Earth when at its farthest point (*aphelion*) than at its nearest point (*perihelion*).

At any particular place on Earth the amount of solar radiation striking the top of the atmosphere also depends on two other orbital parameters. One of these is the tilt of the Earth's axis to the plane of the orbit. This tilt is responsible for seasonal variations in weather and climate: when one pole is tilted toward the Sun that hemisphere experiences summer, and the other winter; half a year later the seasons are reversed when the other pole points toward the Sun. The extreme points are referred to as the *summer* and *winter solstices*; between these are the *spring* and *autumn equinoxes*, when neither pole is directed toward the Sun, which is then directly overhead at the equator at noon. The tilt varies cyclically with a period of about 41,000 years; its value ranges between about 22.1° and 24.5°, with a current value of about 23.5°. The regions of the Earth within 23.5° latitude of the equator, called the *tropics*, experience an overhead Sun twice per year, while the regions within 23.5° latitude of each pole experience a midnight Sun around the time of the summer solstice and no sunlight at all around the winter solstice.

The third orbital parameter which affects the timing and distribution of solar radiation on Earth is the position on the elliptical orbit of the solstices and equinoxes. This is determined by the *precession* (wobble)

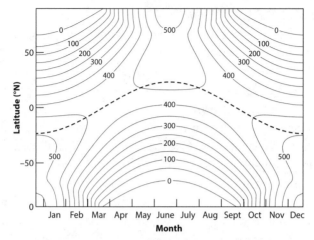

Figure 4.2 Daily average solar radiation (W m^{-2}) at the top of the atmosphere as a function of time of year and latitude. The dashed line indicates the latitude at which the Sun lies overhead at midday.

of the Earth's axis, which varies with periods of about 19,000 and 23,000 years. Its current value is such that the Northern Hemisphere summer solstice is approximately coincident with the orbit's aphelion, so that the Northern Hemisphere receives about 7% less radiation in summer, and 7% more in winter, than the equivalent seasons in the Southern Hemisphere.

The distribution of daily averaged solar radiation incident at the top of the atmosphere throughout the year by latitude is shown in Figure 4.2. Most solar radiation is incident in the tropics, at the subsolar point indicated by the dotted line, but the longer day-lengths in the summer hemisphere mean that, averaged over the day, the greatest amount of radiation is received at middle and high latitudes. Evidence of the ellipticity of the Earth's orbit can be seen in the greater irradiance in the southern summer than the northern.

The solar radiation absorbed at any location is less than the amount incident at the top of the atmosphere because some of it is reflected to space. The amount of reflection depends on the brightness of the surface and the presence of cloud (see the later discussion of the energy budget). The distribution of the absorbed solar radiation is what drives the movements of the atmosphere and oceans. The winds and weather patterns adjust to transport heat away from regions of greater to regions of lesser solar heating. Thus the distribution of solar irradiance over the globe is fundamental in establishing the global circulations of both the atmosphere and oceans as well as regional variations such as the monsoons.

THE EARTH'S RADIATIVE EQUILIBRIUM TEMPERATURE

Fundamentally, the global average temperature of the Earth is determined by a balance between the energy acquired by the absorption of incoming solar radiation and the energy lost to space by the emission of thermal infrared radiation. If the system is in equilibrium, the temperature may be estimated by considering how hot the Earth needs to be to emit a quantity of radiative energy equal to that which it absorbs from the Sun.

Earth viewed as a spherical planet of radius R projects an area πR^2 toward the solar beam and thus absorbs a quantity of radiation equal to $\pi R^2(1 - \alpha)S$, where S is the solar energy flux normal to the solar beam at the Earth's distance from the Sun, and α is the planetary *albedo* (i.e., the fraction of radiation it reflects). If Earth behaves as a perfect emitter (*blackbody*) at a temperature T_e, then it emits thermal radiation to space of $\sigma T_e^{\,4}$ per unit surface area, where σ is the Stefan-Boltzmann constant, and temperature is measured in Kelvin. Balancing the absorbed solar energy with the total emitted energy gives

$$\pi R^2(1 - \alpha)S = 4\pi R^2 \sigma T_e^4$$

Thus

$$\sigma T_e^4 = (1 - \alpha)S/4 = F_s$$

where F_s is the absorbed solar irradiance averaged over the globe, and T_e is the Earth's "equilibrium temperature." Taking $S = 1365$ W m^{-2} (we use this figure rather than the most recent estimate of 1361 W m^{-2} [see Chapter 3] for consistency with other studies cited later) and $\alpha = 0.3$ as values appropriate to the Earth (see Fig. 4.3), we find $F_s = 239$ W m^{-2} and $T_e = 255$ K (or -18 °C). This temperature is much colder than the observed global average surface temperature of the Earth because it corresponds to that of the layer of atmosphere, high above the surface, from which the majority of radiation escapes to space. The surface temperature is greater than T_e owing to the blanketing effect of atmospheric gases, the so-called greenhouse effect.

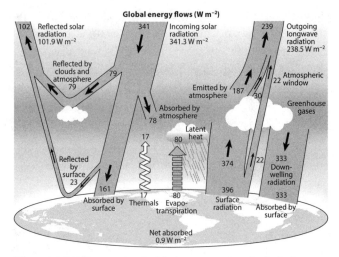

Global energy flows (W m^{-2})

102 Reflected solar radiation 101.9 W m^{-2}

341 Incoming solar radiation 341.3 W m^{-2}

239 Outgoing longwave radiation 238.5 W m^{-2}

Reflected by clouds and atmosphere 79

Atmospheric window 22

Emitted by atmosphere 187 30

Greenhouse gases

Absorbed by atmosphere 78

Latent heat

17 80

Reflected by surface 23

374 22 333 Down-welling radiation

161 17 80 396 333

Absorbed by surface Thermals Evapo-transpiration Surface radiation Absorbed by surface

Net absorbed 0.9 W m^{-2}

Figure 4.3 Earth's annual and global average energy budget, shows fluxes of energy (W m^{-2}) (Adapted from Randall, 2012, fig. 1.3)[20]

EARTH'S ENERGY BUDGET AND THE GREENHOUSE EFFECT

Figure 4.3 gives more details of how radiation is transferred within the atmosphere and at the surface. The value for the incoming radiation, 341 W m^{-2}, is equivalent to a total solar irradiance at the Earth of 1365 W m^{-2} averaged over the globe. Of this about 30% (102 W m^{-2}) is reflected to space by clouds, aerosols, atmospheric molecules, and the surface, with the clouds playing the most important role, so that only 239 W m^{-2} is absorbed by the Earth system. Of the remaining incoming radiation, 23%

(78 W m^{-2}) is absorbed within the atmosphere, leaving 47% (161 W m^{-2}) to reach and heat the surface.

The surface and atmosphere warm owing to the absorption of the solar energy, and so they emit heat in the form of thermal infrared radiation. Figure 4.3 shows that the temperature and emission properties of the surface are such that it emits 396 W m^{-2} of infrared energy. However, only 6% (22 W m^{-2}) of this energy escapes to space; the remainder is absorbed by atmospheric gases and cloud. The atmosphere returns 333 W m^{-2} to the surface. The energy balance at the surface is achieved by nonradiative processes such as evaporation and convection. The radiation balance at the top of the atmosphere is achieved by the 217 W m^{-2} of heat energy emitted to space by the atmosphere itself. Changes to any of the energy flux components affect the equilibrium surface temperature and the climate. Thus variations in total solar irradiance or planetary albedo—which is determined by land cover and cloud and aerosol properties—have an effect on climate, as do changes in the concentrations of any of the gases involved in determining atmospheric transmittances. For solar radiation these include ozone (O_3), water vapor (H_2O), and nitrogen dioxide (NO_2); and for infrared radiation the "greenhouse gases" water vapor, carbon dioxide (CO_2), methane (CH_4), nitrous oxide (N_2O), ozone, and chlorofluorocarbons (CFCs).

While this picture represents the global average situation, it should be remembered that the radiative fluxes vary widely across the globe, with season and cloud cover. For example, satellite measurements show that the absorbed solar radiation over cloud-free oceans in the tropics can exceed the outgoing thermal radiation by more than 120 W m^{-2}, whereas the converse is true over icy surfaces in polar regions, even in summer.

The greenhouse effect is a result of the situation, portrayed in Figure 4.3, that the atmosphere is relatively transparent to solar radiation, thus allowing it to reach and warm the surface, while being absorptive in the infrared, thus trapping the heat energy at low levels. It should be noted that the label "greenhouse" in this context is a misnomer in that garden greenhouses keep warm mainly by inhibiting convection (i.e., by trapping the hot air) rather than by limiting infrared emission. Nevertheless, the terminology has achieved such widespread acceptance that it is retained in this volume.

The greenhouse gas making the greatest contribution to warming the Earth's surface to its habitable value of +14 °C, rather than the −18 °C equilibrium temperature discussed earlier, is water vapor. The next most important gas is CO_2, which, at its current concentration, is contributing about one-third as much as H_2O. The increase in CO_2 and other greenhouse gases from human activity is the main source of the global warming observed over the past century; however, this warming is enhanced by an associated increase in H_2O, because the warmer atmosphere is able to hold more moisture. Changes in cloudiness can either enhance or reduce the initial warming, depending on cloud altitude, location, and microphysical properties.

RADIATIVE FORCING AND
CLIMATE SENSITIVITY

In equilibrium, and on a global annual average, as discussed previously, the net radiative flux at the top of the atmosphere (TOA) is zero. However, if there were a change in solar irradiance, surface albedo, or concentration of a greenhouse gas, then the net TOA flux would not be zero until the climate system adjusted to a new equilibrium. The simplest definition of *radiative forcing* (RF) is this instantaneous change in the TOA flux: if the RF is positive, then there is an increase in energy entering the system (or, equivalently, a decrease in energy leaving the system) and it will tend to warm until the outgoing energy matches the incoming and the net flux is again zero.

The concept of radiative forcing has been found to be a useful tool in analyzing and predicting the response of surface temperature to imposed radiative perturbations. General circulation models of the coupled atmosphere-ocean system (AOGCMs) have found that, approximately, the change in globally averaged equilibrium surface temperature is linearly related to the radiative forcing by a factor, the *climate sensitivity*, which is fairly insensitive to the nature of the perturbation. Two definitions of climate sensitivity are used: The *equilibrium climate sensitivity* (ECS) is the increase in global mean surface temperature in response to a doubling of atmospheric CO_2 after the whole climate system has equilibrated. The *transient climate response* (TCR) is the temperature increase that would occur if CO_2 levels increased by 1% (compounded) per year until they reached the same value (i.e., double the initial). TCR is always smaller than ECS because of the very long timescales required for the oceans to reach equilibrium.

The Intergovernmental Panel on Climate Change (IPCC, 2013)[21] found that ECS is likely to be in the range 1.5–4.5 K, and TCR in the range 1.0–2.5 K. The large ranges represent the spread of values given by different AOGCMs (see the discussion in Chapter 2). (Note that the RF associated with a doubling of the CO_2 concentration is about 4 W m^{-2}, so equilibrium sensitivity can be expressed as the range 0.04–1.13 K W^{-1} m^2). These values give an indication of the uncertainties in climate prediction. It should be noted, however, that for each *particular* GCM the range found using different sources of radiative forcing is much narrower, which suggests that although absolute predictions are subject to some uncertainty, the forecast of the relative effects of different factors is much more robust.

Solar Radiative Forcing of Climate Change

In Chapter 6 we consider potential radiative forcing by increased total solar irradiance in the context of forcing by other factors, specifically greenhouse gases. A common measuring stick is the radiative forcing of approximately 4 W m^{-2} that would ensue from a doubling of the concentration of atmospheric carbon dioxide from its preindustrial value (about 275 parts per million [ppm]);

at current emission rates this concentration is likely to be reached by the year 2050. For the Sun to produce a radiative forcing of ±4 W m^{-2} the total solar irradiance S would have to change by about ±23 W m^{-2} (or ±1.7%), because, as outlined earlier, the global average TOA incoming solar flux F_s is related to S by $F_s = (1 - \alpha) S/4$.

SOLAR RADIATION WITHIN THE EARTH'S ATMOSPHERE

The atmosphere absorbs solar radiation, and while this energy is first used in photodissociation, molecular excitation, or ionization processes, it eventually ends up as molecular kinetic energy—that is, it raises atmospheric temperatures. Although only about 78 W m^{-2} (23%) of the incident solar radiation is absorbed directly within the atmosphere, this energy is fundamental in determining the vertical temperature structure. How much, and where, solar radiation is absorbed depends on the concentrations and physical properties of the atmospheric constituents. The absorbing properties vary considerably with the wavelength of the radiation, so the spectrum of solar radiation is crucial. Figure 4.4 (a) shows a blackbody spectrum at 6000 K, approximately the Sun's radiating temperature, representing the insolation at the top of the Earth's atmosphere. (A more realistic spectrum is presented in the top panel of Fig. 3.2.) Panels (b) and (c) of Figure 4.4 show spectra of atmospheric absorption as the fraction of the incident radiation absorbed at each wavelength. Absorption features due to specific gases are

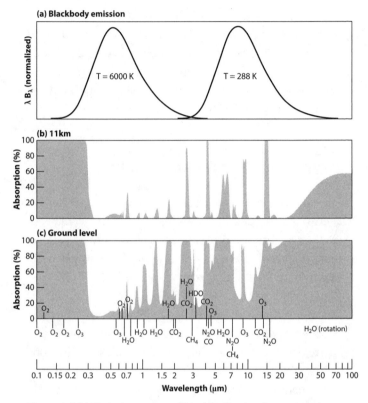

Figure 4.4 (a) Emission spectra of blackbodies (perfect emitters) at the approximate surface temperature of the Sun and of the Earth normalized to show equal areas representing radiative energy flux into and out of the Earth system. (b) Percent absorption of radiation between the top of the atmosphere and 11 km (c) Percent absorption of radiation between the top of the atmosphere and the ground. (Goody and Yung, 1989)[22]

identified: In the UV region, at wavelengths shorter than about 0.3 μm, molecular oxygen (O_2) and ozone (O_3) in the atmosphere absorb all the incident radiation. From the near-UV across the visible region (approximately 0.3–0.7 μm) there is only very weak absorption (mainly by ozone), so that most of the solar radiation at these wavelengths reaches the Earth's surface. At longer wavelengths, into the near-IR region, the narrow absorption bands are due to water vapor (H_2O) and carbon dioxide (CO_2).

The flux of solar radiation at any wavelength reaching a given level in the atmosphere is the product of the incident flux and the fractional transmittance of the path from the top of the atmosphere to that level. The transmittance depends on the integration along the atmospheric path of the amount of radiatively active gas combined with its absorption and/or scattering properties. These can have strong spectral dependency, so that at wavelengths with strong absorption, low values of transmittance are reached at higher altitudes, and vice versa. At wavelengths shorter than 0.2 μm strong absorption by atomic and molecular oxygen and nitrogen means that very little radiation reaches farther down than the stratopause. Radiation in the 0.2–0.3 μm range is absorbed mainly by ozone in the stratosphere. Absorption in the visible and near-IR is much weaker, but because this region is near the peak of the solar spectrum, the energy deposition into the lower atmosphere is significant. The gray curves in Figure 3.2 (a) indicate the typical spectrum of solar radiation reaching the surface.

Solar irradiance varies with solar activity, as discussed in Chapter 5, and it is clear from the preceding discussion that the spectral composition of this variation will be important in determining which parts of the atmosphere will feel the greatest direct effect.

Solar Heating Rate

Most of the absorbed solar radiation is eventually converted into thermal energy, so that the local solar heating rate can be estimated from the convergence of the direct solar flux. Consequently, the heating of the atmosphere by the direct solar beam is directly proportional to the product of the solar flux at that location with the fractional concentration of the absorbing gas and its absorption coefficient.

Figure 4.5 shows solar spectral heating rates calculated in the 200–700 nm region as a function of height in the atmosphere for latitude 57° N in December. Radiation between 200 and 242 nm is absorbed by molecular oxygen, heating the stratopause region and producing the oxygen atoms important in the production of ozone. Across the 200–350 nm region the radiation is responsible for the strong radiative heating in the upper stratosphere and lower mesosphere as well as the photodissociation of ozone. Ozone absorption bands around 440–800 nm are much weaker, but because they absorb broadly across the peak of the solar spectrum, their energy deposition contributes to heating throughout the stratosphere.

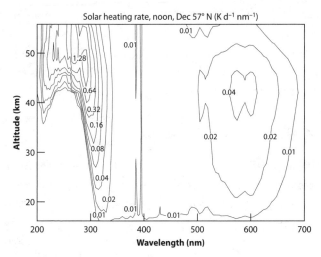

Figure 4.5 Spectrum of solar heating rate (K day^{-1} nm^{-1}) in the ultraviolet and visible regions as a function of wavelength and altitude in the atmosphere calculated for a clear sky at latitude 57° N on December 21 at noon.

If the solar spectral irradiance changes, as discussed in Chapter 5, then—without any impact on composition—the spectral heating rate just varies in proportion to the irradiance. However, if the atmospheric composition also responds to solar variability, then this change will affect both the flux of radiation, F, at any particular height and the heating rate, Q, in a nonlinear fashion. For example, an increase in TOA flux will tend to enhance F and Q, but if it also results in an increase in concentration of an absorber (ozone is the most obvious example), this increase will tend to reduce F at levels below. The sign of the change in F at any altitude depends on the competition between these two factors, namely, the spectral composition of the change in TOA flux and its effects on the photochemistry of the atmosphere. The effect on Q is then a combined effect of the changes in F and the absorber concentration.

SCATTERING OF RADIATION

The absorption of solar radiation, as discussed, is the main source of direct heating of the atmosphere, but processes involving reflection and scattering can have significant impact by redirecting beams. Figure 4.3 shows that 79 W m^{-2} (23%) of the incident solar radiation is reflected to space by clouds and atmospheric gases. In general, the fraction of radiation incident on a material that is deviated by scattering, and also the direction in which it is scattered, depends on the size of the scattering particle. If it is much smaller than the wavelength of the radiation, then the scattering is strongly wavelength dependent: examples are sunlight scattered by air molecules, with the blueness of the sky revealing the stronger scattering of shorter wavelengths; and the scattering of microwave radiation by raindrops, which is used by satellite instruments to detect precipitation. If the particle size and wavelength are of similar magnitude, the situation is complex, but scattering is generally weakly wavelength dependent; examples are sunlight scattered by dust particles, or infrared radiation by cloud droplets. Finally, if the particle is much larger than the wavelength, then geometric optics apply; examples are sunlight scattered by raindrops (rainbows, glories, halos), or infrared radiation by ice crystals.

In areas with a large number of particles the particle size distribution determines the bulk response (within one of the scattering regimes outlined). For example, a cloud forming in a region with a given amount of liquid water available might be composed of a large number of small droplets or a smaller number of larger drops. It transpires that the former situation is more effective at scattering radiation, and such a cloud has a higher reflectivity. Thus if a large number of cloud condensation nuclei (see Chapter 2) are present, the drop size will be smaller, and the albedo of the cloud will increase. This behavior is fundamental to the proposed link between solar activity—through its impact on cosmic rays—and cloud cover, as outlined in Chapter 7.

ATMOSPHERIC TEMPERATURE PROFILES

To balance the absorbed solar energy the atmosphere must lose heat by radiating energy in the thermal infrared. The amount of energy radiated depends on the local temperature and on the IR spectral properties (emissivities) of the atmospheric constituents. Figure 4.4 (a) shows a blackbody spectrum at 288 K, representing the surface temperature of the Earth with a peak at thermal IR wavelengths. The solar and terrestrial spectra barely overlap, allowing us to discuss separately the effects of solar and infrared radiation on climate. Also shown in Figure 4.4 (b) is the atmospheric absorption spectrum; far-IR radiation (wavelengths longer than about 20 μm) is strongly absorbed by water vapor, as is thermal IR, especially at 6.3 and 2.7 μm, where strong H_2O absorption bands appear. Together these features are mainly responsible for the natural greenhouse effect on Earth. Other greenhouse gases that appear in the absorption spectrum include CO_2, with strong bands at 15, 4.3, and 2.7 μm; methane (CH_4) at 3.3 and 7.7 μm; nitrous oxide (N_2O) at 4.5 and 7.8 μm; and O_3 at 9.6 μm. Features due to chlorofluorocarbons appear as well.

Where the atmosphere is optically thin in the IR, such as in the stratosphere, radiant heat energy may be transmitted directly to space, causing local cooling. At lower altitudes, where the atmosphere is optically much thicker, the emitted IR radiation is absorbed and reemitted by neighboring layers. Thus the atmospheric temperature profile is determined by interactions between levels as well by solar heating and direct thermal emission.

In the middle atmosphere (i.e., the upper stratosphere and mesosphere) absorption of solar UV radiation by oxygen and ozone, as described in the previous section, produces the peak in temperature at the stratopause. This heating is counteracted by thermal emission, mainly by carbon dioxide at about 15 μm, but also by ozone at 9.6 μm and water vapor around 6.3 μm. In the lower stratosphere (between approximately 15 and 25 km) heating is due to ozone absorption of visible radiation and also of IR radiation emanating from lower levels. Cooling here is mainly by thermal emission by carbon dioxide.

In the troposphere both IR and solar radiation are largely transferred by water vapor, with the former being dominant. Radiative processes do not, however, determine the temperature profile in this region, because a radiative equilibrium profile would be convectively unstable; that is, a small upward displacement of an air parcel would not produce sufficient adiabatic cooling to make it colder than its environment, and thus it would continue to rise. The temperature profile of the troposphere is therefore limited by convective processes, resulting in an approximately linear decrease of temperature with height (*lapse rate*) of about 7 K km^{-1}. Temperatures are locally warmer than would be the case based on radiative processes alone, so that IR emissions increase, resulting in IR cooling due to tropospheric water vapor rather than the warming which would come about from IR trapping alone. This means that the tropopause marks the region where radiative processes (ozone heating and carbon dioxide cooling) take over from mainly convective processes. Note that an important factor determining the value for the temperature lapse rate is the release of latent heat from the condensation of water vapor into cloud droplets. Thus clouds play an integral part in determining the temperature structure of the lower atmosphere.

Higher up, above the stratopause, the effects of UV absorption by ozone are reduced, and there is a minimum in temperature at the mesopause. Higher still, heating due to the absorption by molecular oxygen of far-UV radiation takes over, and there is a steep increase in temperature in the lower thermosphere.

PHOTOCHEMISTRY

Solar radiation plays a key role in determining chemical composition. Shorter wavelengths penetrate less far into the atmosphere than longer ones, so processes such as ionization which require higher energy (X-ray or far-UV) photons occur only in the upper atmosphere. Near-UV wavelengths, however, are involved in the excitation of atomic energy levels and the photodissociation of molecules throughout the middle and lower atmosphere. The dissociation products are frequently more chemically reactive than the parent molecule, so the presence of solar radiation is fundamental to the chemical processes which take place.

Stratospheric Ozone

In the stratosphere the main chemical reactions determining the concentration of ozone are

$$O_2 + h\nu \, (\lambda < 242 \, \text{nm}) \rightarrow O + O \tag{1}$$

$$O + O_2 + M \rightarrow O_3 + M \tag{2}$$

$$O_3 + h\nu \, (\lambda < 310 \, \text{nm}) \rightarrow O_2 + O \tag{3}$$

$$O + O_3 \rightarrow 2O_2 \tag{4}$$

$$O_3 + X \rightarrow O_2 + XO \tag{5}$$

$$O_2 + X \tag{6}$$

In reaction 1 the term $h\nu$ represents the energy of a photon of radiation at wavelengths less than 242 nm

(= 0.242 μm). This photon has enough energy to dissociate an oxygen molecule (O_2) and is the key step in ozone formation, because the oxygen atoms produced react with additional oxygen molecules to produce ozone molecules, as depicted in the second reaction (the M represents any other air molecule whose presence is necessary to simultaneously conserve momentum and kinetic energy in the combination reaction). The short-wavelength UV radiation becomes depleted as it passes through the atmosphere, which is reflected in the concentration of atomic oxygen (O). The ozone profile does not take a similar form because the reduction in UV is counterbalanced by the need for a three-body collision (reaction 2)—which is more likely at higher pressures (lower altitudes)—so that a peak in O_3 production occurs at around 50 km. Reaction 3 is the photodissociation of O_3 by radiation of wavelength shorter than 310 nm, into O and O_2. However, this dissociation does not represent the primary destruction of O_3, because the O atom produced can quickly recombine with an O_2 molecule. Reaction 4 represents the destruction of O_3 by combination with an O atom to produce two molecules of O_2.

Reactions 5 and 6 represent the destruction of "odd oxygen," that is, O_3 and O, by a catalyst X, which remains unaffected while the oxygen species revert to the diatomic molecule O_2. Possible catalysts include OH, NO, and Cl, with various destruction paths important at different altitudes. Their combined effect is an ozone density profile which peaks near 25 km altitude in equatorial regions.

Ozone production (in terms of the numbers of molecules created per unit volume per unit time) peaks in the low latitudes of the middle stratosphere in summer, where the combination of UV irradiance and atmospheric density is optimum. Observations show, however, that the quantity of ozone above a unit area of the Earth's surface (referred to as the *ozone column amount*) is usually greatest at midlatitudes in winter and spring. This maximum is due to transport by atmospheric circulations which tend to move the ozone away from its source region toward the winter pole and downward. In the lower stratosphere the photochemical lifetime of ozone is much longer, because of the reduced penetration of the radiation which destroys it, and its distribution is determined by transport rather than by photochemical processes. Thus the poleward transport results in an accumulation of ozone in winter high latitudes until the return of sunlight in the spring. In the Northern Hemisphere the peak values occur near the pole, whereas in the south the very strong circumpolar winds tend to restrict the transport resulting, so that the greatest ozone column occurs around 55° S.

The Ozone Hole

This natural minimum in ozone column near the South Pole has deepened considerably during the spring (October) of the last 30 years of the twentieth century into what has become known as the *ozone hole*. Observations and theoretical studies have shown that the depletion occurs mainly in the lower stratosphere and results from catalytic destruction by chlorine (and also bromine). Chlorine reservoir species HCl and $ClONO_2$ are converted into Cl_2 molecules through reactions on the surfaces of polar stratospheric clouds which form in the very cold temperatures present over the South Pole in winter. When the Sun rises in spring the Cl_2 molecules are dissociated into Cl atoms, which can destroy ozone. The catalytic cycle described in the previous section cannot take place because of the very low concentrations of atomic oxygen, but one involving $(ClO)_2$ is possible, and it is this cycle that is mainly responsible for the significant destruction which creates the ozone hole.

The sources of the Cl_2 within the reservoir species are chlorofluorocarbons (CFCs), which were banned under international agreement in 1989. Because of the long atmospheric lifetime of CFCs, and the natural interannual variability of the winter polar atmosphere, the ozone hole is only now slowly filling, according to indications.

Energetic Particles and Chemistry

As outlined in Chapter 3, solar activity is manifest not only in variations of emitted electromagnetic radiation but also in a range of other parameters. One of these is the occurrence and severity of coronal mass ejections resulting in the emission of energetic particles, some of which reach the Earth. The highest-energy particles penetrate well into the stratosphere and affect its chemical composition. Precipitating electrons and solar protons affect the nitrogen oxide budget of the middle atmosphere through ionization and dissociation of nitrogen and oxygen molecules. NO catalytically destroys ozone, as discussed earlier, and reductions in ozone concentration extend down to the middle stratosphere following particularly energetic events. As the solar particles follow the Earth's magnetic field lines these effects have greatest initial impact near the poles, but ozone depletion regions may propagate downward and equatorward over a period of a few weeks. It is interesting to note that the effect of energetic particle events on ozone is the opposite of that of enhanced UV irradiance. As particle events are more likely to occur when the Sun is in an active state, the geographic, altitudinal, and temporal distribution of combined effects on ozone may be complex.

Tropospheric Chemistry

Solar radiation is also fundamental in determining the composition of the troposphere. The daytime chemistry of the troposphere is dominated by the hydroxyl radical ($OH\cdot$) , because its high reactivity leads to the oxidation and chemical conversion of most other trace constituents. The $OH\cdot$ radical is formed when an excited oxygen atom, $O(^1D)$, reacts with water vapor. The source of the $O(^1D)$ is the photodissociation (at wavelengths less than about 310 nm) of ozone; thus the presence of ozone is fundamental to the system. A major source of tropospheric ozone is transport from the stratosphere, but it is also formed through the photolysis (at wavelengths less than 400 nm) of nitrogen dioxide (NO_2), which can be catalytically regenerated. Because OH· is photolytically produced its concentration drops at night, and the dominant oxidant becomes the nitrate radical (NO_3·), which is itself photochemically destroyed during the day.

In this chapter we have seen that solar output influences the temperature of the atmosphere through direct heating and affects its composition through photochemical processes. In the next chapter we consider solar activity and how changes in the Sun are associated with variations in its output of radiation and particles.

5   SOLAR VARIABILITY

Having addressed the underlying properties of the solar atmosphere in Chapter 3, we now discuss their temporal variation. The most important one is the reverse in polarity of the solar magnetic field roughly every 11 years, so a complete solar cycle occurs approximately every 22 years. The reason for this reversal must lie in the dynamo process operating at the tachocline, though a complete explanation is still awaited. This lack of understanding also underlies our inability to predict, on a first-principles basis, events like the recent deep and prolonged solar minimum. Note also the use of the word "roughly" to describe the time of the cycle: the length of any particular cycle does vary from precisely 11 years.

VARIATIONS IN TSI ACROSS A SOLAR CYCLE

From the viewpoint of radiative input to the Earth, we are interested in the variability of visible and UV radiation both across a typical solar cycle and, longer term, over many cycles. We emphasized in Chapter 3 that calculation of TSI requires space-based measurements owing to the absorption of, in particular, UV radiation by the atmosphere. Figure 5.1 shows the daily averaged TSI from a number of spacecraft since the start of regular space measurements.

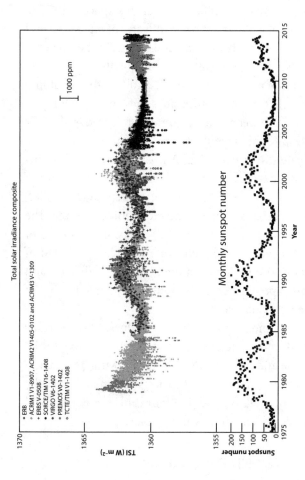

Figure 5.1 The daily averaged TSI from a range of spacecraft between 1980 and the present (the acronyms refer to the various spacecraft or instruments). Note the overlap between all the data sets which permits cross-calibration. A composite TSI can be constructed from these, giving the present value of the TSI (1361 W m^{-2}). (Figure courtesy of G. Kopp, http://spot.colorado.edu/~koppg/TSI/)

The first systematic measurements were made in the early 1980s and continue to the present day. Figure 5.1 presents TSI data from various space missions. The original datasets showed offsets, of up to several W m^{-2}, resulting from the differences in the absolute TSI measured by the various instruments. Fortunately, however, the records overlap one another, thus enabling the construction of a composite series, and in the figure the data have been inter-calibrated to remove the offsets. Despite the large fluctuations from day to day, the composite shows overall agreement in the trends of increase and decrease over each solar cycle, with values of TSI around 1 W m^{-2} higher at solar maxima (1980/1981, 1991, 2001/2002) relative to the intervening minima. The prolonged low value of TSI at the recent minimum of 2008/2010 can also be seen. Note that change in TSI of well under 0.1% is small in the context of other natural and anthropogenic contributions to radiative forcing of climate (see Chapter 6).

The remainder of this chapter discusses how various solar and interplanetary parameters vary relative to the solar cycle and how these are correlated with TSI. Figure 5.2 shows summary plots of various parameters over the past four decades (a) and past four centuries (b) and will be referred to hereafter as appropriate.

SUNSPOTS AND FACULAE

The most familiar manifestation of the solar cycle is the variation in the sunspot number (top panel of Fig. 5.2 (a) and fifth panel of Fig. 5.2 (b)), with the 11-year periodicity evident. The dimensionless (Wolf) number is calculated as $R \approx 10N + n$, where N is the number of sunspot groups and n the number of individual spots. Two points should be noted. The first is the *Maunder Minimum* from 1640 to 1710. (Of course, the lack of visible spots during the Maunder Minimum should not be construed as meaning that the Sun was entirely magnetically inactive.) The second is the variation in the maximum of the sunspot number since roughly 1750 by a factor of up to 3 (contrast 1800 during the *Dalton Minimum* and 1960).

In Chapter 3 we noted that faculae are an important (brightening) effect offsetting sunspot darkening in determining the TSI. Figure 5.3 shows the estimated change in TSI due to faculae (upper curve) and sunspots (lower curve). The facular brightening is smoother, because it arises from many small sources as opposed to fewer large spots. In fact, the facular brightening is greater than the sunspot darkening, so that the TSI increases at the maximum of the activity cycle. At minimum activity conditions, neither faculae nor sunspots matter very much.

UV, EUV, AND X-RAYS

The behavior of the outer solar atmosphere through a solar cycle correlates well with the sunspot number. The most intense sites of radiation are active regions (ARs). These are associated with sunspot groups—hence stronger magnetic fields—and are readily evident on full-Sun images of chromosphere and corona. Since it is believed

(a)

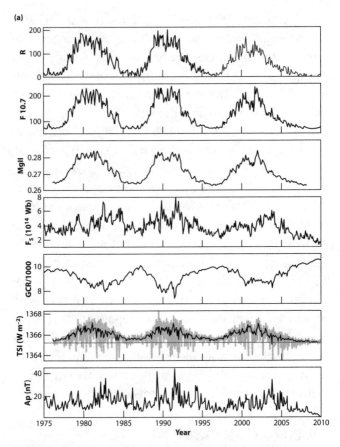

Figure 5.2 Variation of various solar and terrestrial quantities over the past four decades (a) and past four centuries (b). (a) Panels show the sunspot number; F10.7 radio flux; Mg II core-to-wing ratio; interplanetary magnetic flux (1 Wb = 1 T m^{-2}); galactic cosmic rays (GCRs) (neutron counts per minute) recorded in McMurdo, Antarctica; TSI; and the auroral Ap index. All except GCRs are positively correlated, though some signals are stronger than others.

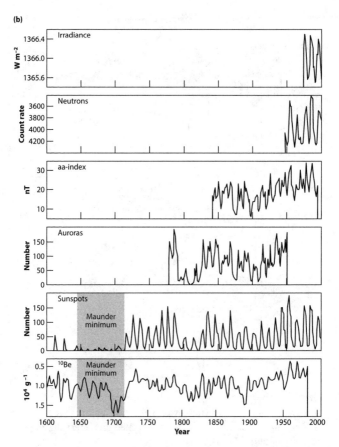

Figure 5.2 (*continued*) (b) Panels show TSI; GCR neutrons recorded at Climax, Colorado; the auroral aa index; number of auroras; sunspot number; and ^{10}Be isotope concentration. The time series are plotted only for reliable data. (Gray et al., 2010)[23]

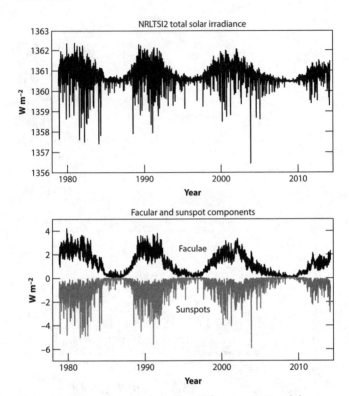

Figure 5.3 The upper panel shows the TSI between 1979 and the present. The lower panel shows the change in TSI due to facular brightening and sunspot darkening over the same interval. (Figure courtesy of Judith Lean)[24]

that the presence of chromospheric and coronal activity depends on the magnetic field strength, it is reasonable to expect this correlation.

At cycle maximum, the very intense active region emission occupies a significant fraction (few percent) of the solar surface. At cycle minimum, the active regions almost vanish. These active regions can exist for well over a solar rotation and are a quasi-steady source of radiation in EUV, UV, and X-rays. As seen in Table 3.2, the power output from ARs is up to two orders of magnitude higher than that from the quiet Sun. Outside ARs, the temperature of the quiet Sun does not exceed 2–3 MK, with peak emission at just over 1 MK. In contrast, ARs can have a peak emission at a few MK and may have nonflaring components up to 10 MK, though with small emission. Interpretation of this last phenomenon is still controversial.

Although the level of radiation at wavelengths shorter than the visible is very important, much of this is absorbed before reaching the ground, so that continual monitoring from space is highly desirable. However, this relies on continuity of space observations and proper calibration. In lieu of space observations, ground-based proxies are used. The advantage of proxies is that they can provide a long time series and are cheap. The disadvantage is that they *are* proxies and often involve the interpretation and parameterization of rather complicated physics

The flux of radio emission at 10.7 cm wavelength (or a frequency of 2.8 GHz) has been documented for many decades over the entire solar disk. The choice of frequency is historical, dating to the late 1940s. The physics behind these measurements is a mixture of thermal bremsstrahlung and thermal gyroresonance emission.[25, 26] Thus the emission is enhanced in regions of high density and strong magnetic field. The second panel of Figure 5.2

(a) shows F10.7 values over the last few cycles, in solar flux units (1 SFU = 10^{-22} W m^{-2} Hz^{-1}). The power in this radiation is extremely small. Short-term EUV and X-ray transients such as flares with a rise time of a few minutes and decay lasting up to a few hours also persist throughout the cycle, though on the whole, large flares vanish at the time of solar minimum. A second proxy for solar UV and EUV irradiance is the *Mg II core-to-wing ratio* calculated from the Mg II ion emission spectrum centered around 280 nm (Figure 5.2 (a), third panel). It is argued that the "core" of the spectrum at 280 nm is generated in the chromosphere, and the "wings" in the photosphere, so the ratio is expected to increase as magnetic activity increases. Indeed, this is what is found. Spectral parameterizations or other modeling of these quantities can be used as inputs to atmospheric models.

THE SOLAR WIND, INTERPLANETARY MAGNETIC FIELD (IMF), CORONAL MASS EJECTIONS, AND AURORAS

We discussed in Chapter 3 the supersonic expansion of the solar corona as the solar wind. The evolution of the wind over the solar cycle reveals much about the global evolution of the solar magnetic field, as well as being of importance for the interplanetary medium. Results from the out-of-the-ecliptic Ulysses mission between 1992 and 2005 revealed the behavior of the solar wind at the minimum and maximum of the solar cycle. At the minimum of the cycle, there are two distinct regions of wind. At high latitudes there is a high-speed (700 km s^{-1}) wind, with an abrupt transition to lower speeds at around 30° N and S. The magnetic field is unipolar in these high-speed regions. At solar maximum the wind is less well organized. The high-speed unipolar regions vanish, and a lower-speed wind remains with many large-scale structures embedded. These results reflect the transition through the solar cycle as the magnetic field reverses polarity.[27]

The interplanetary magnetic field (IMF) thus exhibits changes across a solar cycle that are of great consequence in determining the galactic cosmic ray (GCR) flux (see next section). Lockwood and collaborators have documented the variations in the "open" solar wind magnetic flux, as shown in the fourth panel of Figure 5.2 (a). These data (in webers, or T m^{-2}) were calculated from the radial component of the IMF at 1 AU using archival data sets over the space age. For a solar wind magnetic field at the Earth of a few nanotesla, typical magnetic flux values of a few 10^{14} Wb arise. The variations over a solar cycle are perhaps not quite as clear cut as the other quantities we have discussed thus far but are nonetheless present.

Interestingly, in the last decade the value of this open flux fell to below 2×10^{14} Wb—comparable to the lowest values a century ago. How can we make such statements in the absence of actual IMF data prior to the space age? The answer is that reconstruction techniques are used to extend time series back to times before actual data could be obtained. For example, Lockwood and Stamper[28] reconstructed the solar open magnetic flux from 1870 until

2000 using geomagnetic data such as the auroral Ap index shown in the last panel of Figure 5.2 (a) and discussed shortly. Thus the radial flux is determined from geomagnetic data prior to the space age and from in situ measurements thereafter.

The solar wind comprises important large-scale structures—in particular, coronal mass ejections (CMEs). The number of CMEs tends to peak on the downward side of activity as the Sun expels the old polarity field. We return to CMEs in Chapter 8, when we discuss space weather and geomagnetic activity, but here we introduce another observed consequence of the solar cycle—auroral activity, which has been measured in recent years by various auroral indexes. The seventh panel of Figure 5.2 (a) shows the Ap index, and the third panel of Figure 5.2 (b) the aa index. The *Ap index* (in units of nanotesla) is a measure of geomagnetic activity (distortions of the terrestrial magnetic field due to induced currents in the ionosphere and magnetosphere [see Chapter 8]) observed at a number of sites. It correlates with other indicators of solar activity. The associated *aa index* (also in units of nanotesla) has a longer history but uses fewer observing stations than Ap. Both show cyclical behavior, with peak values on the downside of the cycle.

A time series dating to the late eighteenth century tracks the number of auroras (fourth panel of Figure 5.2 (b)). While auroras have been seen for centuries, those at low latitudes have been tabulated only in more recent years. These are important because low-latitude auroras occur only during the more significant geomagnetic disturbances. Unfortunately, this time series terminated in 1950. Clearly, there is a lack of precision in what is meant by "number of auroras." However, although these are not optimal data, they plainly show evidence of the 11-year cycle.

GALACTIC COSMIC RAYS AND COSMOGENIC ISOTOPES

GCRs are charged particles with energies of order hundreds of mega-electron volts per nucleon. They are believed by some to be important for the formation of clouds, as outlined in Chapter 7. Far more certain is that they can be responsible for event upsets on spacecraft (see Chapter 8). The flux of GCRs has been measured in Oulu, Finland, for the last five decades and at other sites such as McMurdo, Antarctica, and Climax, Colorado, for comparable durations. Figure 5.2 shows the overall modulation of the McMurdo ((a), fifth panel) and Climax ((b), second panel) data over the cycle. The large flux during minimum activity indicates that it is easier for the GCRs to travel through the solar wind to the Earth at these times. Note also the large flux at the most recent solar minimum. The small-scale spiky structure in Figure 5.2 (a) is due to individual CMEs around and shortly after the maximum of the solar cycle.

Solar wind turbulence plays an important role in this GCR modulation. A class of turbulence driven by magnetic waves (*Alfvén waves*) is prevalent in the solar wind. In situ observations suggest that these originate at the

Sun and form a so-called turbulent cascade in the solar wind, and the wave power eventually heats the particles. This turbulence is important as a fundamental science problem but is also significant because of its interaction with the GCRs. At the minimum of the solar cycle, the turbulence is weaker, so GCRs undergo fewer interactions with the waves. The opposite is true at the cycle maximum. This pattern accounts for the inverse correlation of GCRs with the other solar cycle parameters.

Evidence for the level of GCR flux in more distant times (see Figure 5.2 (b), sixth panel) can be obtained by measuring the cosmogenic isotope ^{10}Be, which is a spallation product of the interaction of GCRs with atmospheric gases. The levels of ^{10}Be in ice cores, and ^{14}C in trees, are evidence of the solar cycle. With its long half-life of 1.39×10^6 years ^{10}Be can be used to study very long term variations in solar activity, as there is overall agreement between ^{10}Be levels and the "upside-down" sunspot number. There is less ^{10}Be at times of high activity and vice versa. As noted earlier, the CGR flux is responsible for production of ^{10}Be, so modulation of the GCR affects levels of ^{10}Be.

EXTRAPOLATING TSI TO THE PAST

An important question is, given the self-evident correlation between TSI and other quantities over the past three decades, can the TSI be extrapolated to the past (and into the future) using the existing data and other, longer time series? Likewise, can the GCR flux be reconstructed (or predicted) based on IMF measurements and cosmogenic

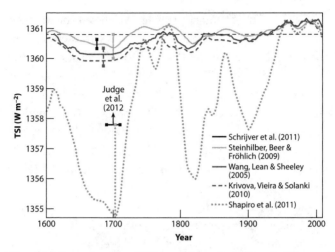

Figure 5.4 Reconstruction of TSI based on various data sets. The predictions of different workers are shown. Most predictions use sunspot numbers or a variation thereof, while Lockwood and Stamper use the relation between open flux and TSI. (Gray et al, 2010)[29]

isotope measurements? This is a process that should send out strong cautionary signals, but it also holds out the possibility of great rewards. It must be seriously considered whether just 30 years of information for TSI, 60 years for F10.7, and so on, can capture all previous conditions, especially extreme events such as the Maunder and Dalton Minima; *caveat lector*.

Figure 5.4 shows TSI reconstruction based on sunspot numbers and gives some idea of the problems at hand. One is that the variation of TSI due to sunspots also involves the level of brightening due to faculae, and measurements of faculae are not available prior to 1874.

Thus the sunspot index prior to 1874 has to be adapted to account for faculae. Another issue is parameterization of the link between the (known) sunspot value and the (calculated) TSI. Solanki and Fligge[30] use a simple quadratic fit, Lean et al.[31] use a more complicated parameterization, and Krivova et al.[32] use a "physics-based model." The reader can investigate the difference between the various approaches. Confidence in any of these depends on one's belief in the parameterization or modeling of the input parameters.

Longer-term reconstructions rely on other parameterizations, in particular between cosmogenic isotopes and climate indicators. Since levels of these isotopes are due to the effects of GCRs, understanding how the isotope parameters relate to the GCR flux and in turn how that is related to TSI and other solar parameters is the key point. That there is an anticorrelation between GCRs and TSI (Fig. 5.2) is well established, and the work of, for example, Steinhilber et al.[33] shows the sort of reconstruction that is possible. They use correlations between TSI and open magnetic flux established from space data in the last 50 years, and a relationship between cosmogenic isotopes and magnetic flux, to calculate changes in TSI over several thousand years. Another approach, adopted by Shapiro et al.,[34] assumes that the minimum state of the quiet Sun in time corresponds to the observed quietest area on the present Sun, suggesting much larger variations in TSI (>6 W m^{-2} from the Maunder Minimum to the present) than proposed by other reconstructions and is included in Figure 5.4. All these techniques remain to be validated.

THE 2008/2010 MINIMUM:
. . . AND THE FUTURE

To conclude this chapter, we discuss the recent "great solar minimum" of 2008/2010. The depth and duration of this has given rise to much conjecture that the Sun may be on a track of declining solar activity and even that a recurrence of something like the Maunder Minimum may be approaching. Figure 5.2 (a) demonstrates this hypothesis. Note (a) the duration for which the sunspot number and F10.7 flux bottomed out, (b) the fact that at this minimum there was a similar level of F10.7 flux, indicating continued magnetic activity, (c) the rise in the peak of the neutron count (and hence GCRs) well above previous maxima, and (d) the slow start of a recovery into the next cycle.

It has been argued by some that this deep minimum is part of a trend to lower activity extending over recent cycles. Also, the reconstruction techniques discussed earlier can be applied to the future. For example, Lockwood et al.[35] note that the probability of Maunder Minimum conditions within 40 years is 5%; likewise, the chance that present levels of activity will be maintained is also 5%, with a range of activity levels between these two extremes, suggesting lower solar activity for the foreseeable future. Although any such decrease in TSI, and its impact on global surface temperature, is likely to be small (see the discussion of the temperature response to the Maunder Minimum in Chapter 6), the decline in solar activity may have an influence on regional climate (see Chapter 7).

Predictions of neutron production suggest a GCR level at least as high as at the last grand minimum—and probably higher—which has clear implications for space-based assets and high-flying aircraft, as discussed in Chapter 8.

6 SOLAR SIGNALS IN SURFACE CLIMATE

In Chapter 5 we considered how the Sun varies in terms of its emissions of radiation and particles. In this chapter and the next we look at how these changes might be associated with variations in weather and climate on Earth.

Investigations of climate variability and climate change depend crucially on the existence, length, and quality of meteorological records. Ideally, records would consist of long time series of measurements made by well-calibrated instruments densely situated across the globe. In practice, of course, this ideal cannot be met. Measurements with global coverage have been made only since the start of the satellite era, in the late 1970s, and records from meteorological instruments running for more than a couple of centuries are available from only a few locations in Europe. For longer periods, and in remote regions, records have to be reconstructed from indirect indicators of climate known as *proxy data*. These proxies provide information about weather conditions at a particular location through records of a physical, biological, or chemical response to the contemporaneous temperature or humidity. Some proxy data sets provide information dating to hundreds of thousands of years ago, which make them particularly suitable for analyzing long-term variations in climate and their relation to solar activity.

One well-established technique for providing proxy climate data is *dendrochronology*, or the study of the successive annual growth rings of trees (which may be analyzed living or dead). It has been found that trees from any particular area show the same pattern of broad and narrow rings corresponding to the weather conditions under which they grew each year. Thus samples from old trees can be used to give a time series of these conditions. Felled logs can similarly be used to provide information about ancient times, providing it is possible to date them. This is usually accomplished by matching overlapping patterns of rings from other trees. One complication that arises with the interpretation of tree rings is that the annual growth of the rings depends on a number of meteorological variables integrated over more than a year, so that the dominant factor determining growth varies with location and type of tree. At high latitudes the major controlling factor is likely to be summer temperature, but at lower latitudes humidity may play a greater role.

Much longer records of temperature have been derived from analysis of oxygen or hydrogen isotopes in ice cores, in particular those obtained from Greenland and Antarctica. The ratio of the concentrations of different isotopes in the water molecules is determined by the rate of evaporation of water from tropical oceans as well as the rate of precipitation of snow over the polar ice caps. Both these factors are dependent on temperature, so greater proportions of the heavy isotopes are deposited during periods of higher global temperatures. As each year's accumulation of snow settles, the layers below become compacted, so that at depths corresponding to an age of more than 800 years it becomes difficult to date the layers precisely. Nevertheless, variations on timescales of more than a decade have been extracted dating back more than hundreds of thousands of years.

An individual proxy record does not give a precise measure of a particular climate parameter and needs to be calibrated by comparison with a contemporaneous instrumental record, as available. The choice of proxy data and of calibration period allows for considerable uncertainty, which is further increased by the treatment of spatial variability and seasonal biases. The result is a sometimes wide divergence between derived records. No individual method is foolproof, and recently there has been increased effort to produce multiproxy records. These probably provide more robust series and also give an indication of the uncertainty in the derived paleo record.

PALEOCLIMATE TEMPERATURE RECORDS

A record of the deuterium ratio, representing temperature, in an ice core retrieved from Vostok in East Antarctica is shown in Figure 6.1. The roughly 100,000-year periodicity of the transitions from glacial to warm epochs is clear and suggests a relationship with the variations in eccentricity of the Earth's orbit around the Sun (see the discussion in Chapter 4), although this variation

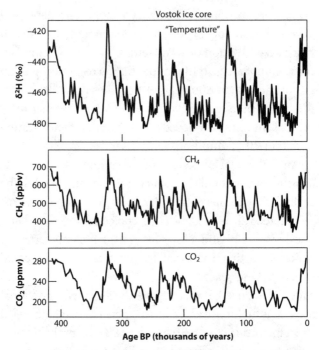

Figure 6.1 Records derived from an ice core taken from Vostok, East Antarctica, showing variations in deuterium ratio (representing temperature) and the concentrations of methane and carbon dioxide over at least 400,000 years. (Stauffer, 2000)[36]

does not explain the steep transitions from cold to warm. Evidence of very long term temperature variations can also be obtained from ocean sediments. The skeletons of calciferous plankton make up a large proportion of the sediments at the bottom of the deep oceans, and the oxygen isotope ratio within these is determined by the temperature of the upper ocean at the time when the living plankton absorbed carbon dioxide. The sediment accumulates slowly, at a rate of perhaps 1 m every 40,000 years, so that changes over periods of less than about 1000 years are not detectable, but ice age cycles every 100,000 years are clearly portrayed.

Figure 6.1 also presents the concentrations of methane and carbon dioxide preserved in the ice core, showing a strong correlation between these and temperature. (Note that neither concentration is nearly as high as the present-day values of around 1893 ppbv [parts per billion by volume; Northern Hemisphere]/1762 ppbv [Southern Hemisphere] CH_4 in 2012, and 397 ppmv CO_2 in 2013). One theory proposed to account for these variations suggests that when the tilt of the Earth's axis is small, the summers are cooler at high latitudes, and thus less of the previous winter's buildup of ice melts, precipitating an ice age. At larger tilts the warming of southern high latitudes caused by the orbital variations is amplified by the release of CO_2 from the southern oceans and further amplified through a reduction in albedo resulting from the melting of Northern Hemisphere ice sheets. Such positive feedback mechanisms might explain the sharp increases in temperature seen in the record. (Note that such increases will likely occur in response to any other warming effect, such as increasing concentrations of anthropogenic greenhouse gases.) Thus cycles of glacial/interglacial periods are probably related to changes in insolation owing to variations in orbital geometry.

This mechanism for ice ages does not involve any variations in intrinsic solar activity or output of radiation. Across the Holocene (the period of about 11,700 years since the last Ice Age), however, isotope records from lake and marine sediments, glaciers, and stalagmites provide evidence that solar grand maxima/minima have affected climate. For example, in the North Atlantic, icebergs resulting from the flow of Greenland glaciers into the sea carry minerals which uniquely define their source. During colder periods the ice is able to raft farther south, where it melts, depositing the minerals. Ice cores and ocean sediments may also preserve, in isotopes such as ^{10}Be and ^{14}C, information on cosmic ray flux and thus solar activity. Consequently, simultaneous records of climate and solar activity may be retrieved from a single sediment or ice core record. The example in Figure 6.2 shows fluctuations on the 1000-year timescale well correlated between the two records, suggesting a long-term solar influence on climate. These ice-rafting events correlate with climate extremes measured in other parts of the world, such as weak events of the Asian monsoon, as indicated by stalagmite records from Oman. A caveat regarding all these studies is that they rely on the dating, which is complex and not always precise.

Proxy temperature data over the past millennium have been collated from a wide range of sources across the globe, including tree rings, ocean and lake plankton and pollen, coral, ice cores, and glaciers, to provide global (and hemispheric) average surface temperature and precipitation records. A composite of published temperature estimates, produced by using different data/

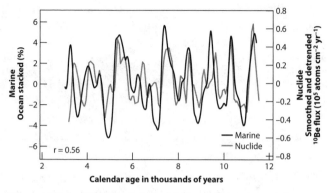

Figure 6.2 Records of ^{10}Be and ice-rafted minerals extracted from ocean sediments in the North Atlantic. (Bond et al., 2001)[37]

techniques, is presented in Figure 6.3, which also shows the uncertainty range of the estimates.

Most of these records show relatively warm values over the period between about 950 and 1250 (sometimes referred to as the *Medieval Climate Anomaly*, MCA) and somewhat cooler values during the sixteenth to nineteenth centuries (sometimes referred to as the *Little Ice Age*, LIA). It has frequently been remarked that the Spörer, Maunder, and Dalton sunspot minima occurred during the LIA, leading to speculation that solar activity might have been the cause of the cooler temperatures. Care needs to be taken in such interpretation, however, as other factors might have contributed. For example, the higher levels of volcanism prevalent during the seventeenth century would also have introduced a cooling tendency owing to reflection

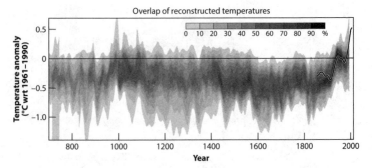

Figure 6.3 Variations in Northern Hemisphere surface temperature over the past millennium compiled from 12 estimates derived from proxy data; the shading indicates the overlap among the uncertainty ranges of the individual records. The black curve indicates the instrumental record since 1850. All values are shown relative to the 1961–1990 average (Based on Figure 6.10c of IPCC AR4)[38]

of the Sun's radiation to space by a veil of particles injected into the stratosphere.

The black curve in Figure 6.3 also represents instrumental measurements of surface temperature compiled to produce a hemispheric average dating to 1860. Much of the current concern about global warming stems from the obvious rise over the twentieth century. Other records suggesting that the climate has been changing over the past century include the retreat of mountain glaciers, rise in sea level, thinning of Arctic ice sheets, and an increased frequency of extreme precipitation events. A key concern of contemporary climate science is to attribute cause(s) to this warming, including ascertaining the role of the Sun.

FACTORS INFLUENCING GLOBAL SURFACE TEMPERATURE

A variety of techniques can be used to assign contributions of individual factors to observed variations in global mean temperature. Several of these are reviewed in the appendix. The results of one attempt using multiple linear regression to analyze temperature over the period 1889–2006 are presented in Figure 6.4. Here the effects of human activity (due to emission of greenhouse gases and particulates), ENSO (El Niño–Southern Oscillation; see Chapter 3), and volcanic eruptions are assessed alongside solar variability. The top panel shows the raw temperature time series in gray. The four lower panels show the contributions inferred to be associated with each of the forcing factors. Each panel shows the forcing index on the right-hand axis and the associated temperature signal on the left-hand axis. Finally, the black curve in the top panel shows the temperature series reconstructed from the individual components; comparison with the gray curve reveals the goodness of fit. These data suggest that the Sun may have introduced an overall global warming (disregarding 11-year cycle modulation) of approximately 0.07 °C, although most of that warming was before about 1960. Over the whole period the temperature increased by about 1 °C, so the fractional contribution to global warming that can be ascribed to the Sun over the last century is 7%. However, this conclusion depends fundamentally on the assumed temporal variation of the solar forcing, and as discussed

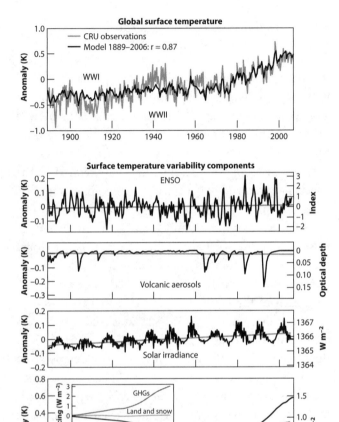

Figure 6.4 Global monthly mean temperature record (black) and reconstruction from multiple regression analysis (gray). The regression contributions are shown for (a) ENSO, (b) volcanic, solar, and anthropogenic (greenhouse gases and particulates) influences (at appropriate lags). (Lean and Rind, 2008)[39]

in Chapter 5, there are several alternatives. The index of solar variability used in Figure 6.4 has a small long-term trend (relative to the 11-year cycle magnitude); the use of other solar indexes might produce a larger signal of temperature increase before midcentury and a better match between observations and regression model in the top panel. Crucially, however, it is not possible to reproduce the global warming of recent decades without including anthropogenic effects, and this conclusion is confirmed by the use of more sophisticated nonlinear statistical techniques.

Another approach uses estimates of radiative forcing (RF; see Chapter 4) to indicate potential global mean temperature change. A 1 W m^{-2} increase in TSI implies a RF of 0.175 W m^{-2} and an increase in global mean surface temperature at equilibrium (using a climate sensitivity of 0.8 K W^{-1} m^{-2}) of about 0.14 K. Thus the implication of the different TSI series presented in Figure 5.4 becomes clear in assessing the role of the Sun in determining the temperature difference between the LIA and the present. That figure shows a range of values for the difference in TSI from the seventeenth century to the present of about 1–3 W m^{-2} (although other published results show even greater disparity). This range of values implies a solar radiative forcing of 0.17–0.50 W m^2 and thus a corresponding solar-induced increase in global average temperature of 0.13–0.4 K since that date, which can be compared with the approximately 1 K increase observed.

TSI reconstructions may also be input to climate models to simulate the role of the Sun in climate history.

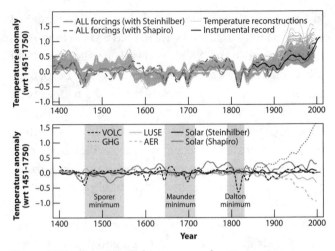

Figure 6.5 Northern Hemisphere mean surface temperature anomalies from observations and model simulations. Upper panel, thin gray lines: Ensembles of reconstructions derived from proxy indicators; black line: instrumental record since 1850; thick solid gray line: mean of model simulations using all forcings (see text for details) with Steinhilber TSI; thick dashed gray line: as solid gray line but using Shapiro TSI (see Figure 5.4 for TSI time series). Lower panel: Simulated contribution of each forcing factor, as identified in the legend (VOLC: volcanic aerosol; GHG: greenhouse gases; LUSE: land use; AER: other aerosol) (Schurer et al., 2013)[40] (Copyright © 2013, Rights Managed by Nature Publishing Group)

The example presented in Figure 6.5 shows the Northern Hemisphere mean surface temperature over the past 600 years from a GCM forced with greenhouse gases, tropospheric aerosols (sulfate, dust, and soot particles), stratospheric (volcanic) aerosols, changes in land use, as well as two different estimates of TSI. The TSI data set due to Steinhilber contains a 1 W m^{-2} overall increase since the Maunder Minimum, and the one due to Shapiro a 6 W m^{-2} increase over the same period. In the upper panel of Figure 6.5 the thin gray lines show a range of reconstructions of the temperature record from proxy measurements; the black line is the instrumental record. The climate model predicted values for all forcings with the Steinhilber TSI series align fairly well, within experimental uncertainties, with the observed/proxy record. The Shapiro reconstruction shows greater warming over the latter half of the twentieth century, but it also extends temperatures outside the range of observed values across much of the 600-year period. In the lower panel, model runs with individual forcings are shown by the various lines identified in the figure caption and legend. The TSI contribution to variations in the temperature anomaly is small throughout the period when the Steinhilber data are used. We noted in Chapter 5 that the Shapiro TSI reconstruction, with its large variations, is an outlier relative to most others, and the larger- amplitude temperature deviations modeled with this TSI series are clear in the figure.

From these reconstructions we conclude that while natural factors—particularly volcanism—are likely to have contributed to variations in temperature over the past 600 years, they cannot account for the sharp warming since about 1970. Studies using optimal estimation approaches (see the appendix) to attribute causes to global temperature change since 1750 come to very similar conclusions without any a priori assumptions on the magnitudes of the forcing factors.[41]

Over the most recent 15 years global mean surface temperatures did not rise as fast as they did during previous decades (see the top panel of Fig. 6.4), nor as fast as predicted by most GCMs for that period. The reasons for this "hiatus" in global warming are currently the subject of intensive study, and variations in the Sun have been proposed as a possible reason. Certainly, overall solar activity was low over that period relative to the previous few decades, but observed changes in TSI were not large enough to compensate for the radiative forcing from greenhouse gases. There may be a small solar component in the recent slowing of the warming trend, as well as a contribution from a number of small volcanic eruptions, but, as demonstrated by similar periods during the century-long data set, it is likely that it represents mainly natural internal variability and a redistribution of energy within the climate system.

It has further been suggested that if the Sun is currently in a state of overall declining activity (see Chapter 5), then the concomitant negative radiative forcing might compensate for the global warming likely to take place in response to greenhouse gas increases on longer timescales. Hypothetically, if the Sun were to enter another grand minimum such as the Maunder event within the next 50–100 years this (by the arguments presented in the preceding discussion) would produce a maximum global cooling of 0.1–0.4 K as compared with the anticipated warming due to unregulated carbon dioxide emissions of 3–4.5 K in the same period. An experiment with an AOGCM[42] assuming a reduction of 0.25% in TSI for a 50-year period confirms this conclusion, showing that such solar behavior would temporarily slow down but not stop global warming. Thus, to rely on such cooling to counter human-produced global warming not only would be a risky strategy, given uncertainties in predicting solar activity, but would also likely produce only a small, temporary compensation.

REGIONAL EFFECTS

Both the statistical and modeling approaches discussed demonstrate that signals of solar variability can be detected in records of the global (or hemispheric) average surface temperature. More detailed analyses suggest that the response is not spatially uniform but that certain parts of the world may experience greater or, indeed, opposite effects. These signals are more difficult to confirm, because the natural "noise" in the time series is not reduced by spatial averaging, but the responses found suggest that a solar signal can be detected in some of the natural modes of variability in the climate system (see Chapter 3).

Solar signals have been identified in meteorological records from many stations across the globe. For example, signals of period around 11 years have been identified from power spectrum analysis of temperature records from stations across the United States, with positive correlations to the east side of the Rockies and negative to the west.[43] In Europe the positions of the storm tracks crossing the North Atlantic have been found to shift north and south with solar activity.[44] An 11-year periodicity has also been identified in cyclones in the tropical Atlantic.[45] There are many other examples. The statistical robustness of these studies is not always well demonstrated, and sometimes the signal changes in and out of phase with solar activity (for a very good discussion see the review by Hoyt and Schatten[46]), making the task of identifying physical causes for the statistical relationships even more challenging.

Optimal estimation techniques (see the appendix) have been used to explicate the processes involved in the statistical analysis. For example, using an energy balance model to generate a distribution over the globe of the response in surface temperature, together with noise estimates from long runs of AOGCMs, Stevens and North[47] identified a small (maximum a few hundredths of a degree Celsius over land in summer) solar signal in the data.

Polar Regions

The Holocene proxy temperature records across the Northern Hemisphere show strong regional variations in the solar response, including an NAO-like pattern (North Atlantic Oscillation; see Chapter 2), such that when the Sun was less active, the climate was characterized in an NAO-negative phase more frequently than normal. Thus during the Sun's Maunder Minimum the surface temperatures were typically cooler in eastern North America and western Europe, and warmer in Greenland and central Asia.[48] The cooler temperatures are consistent with the public experience in Europe, where the longer period including this time was named the Little Ice Age.

There is some evidence that generally higher solar activity during the Medieval Climate Anomaly coincided with warmer temperatures in western Europe and eastern North America, although this result is less robust than the converse found in the Maunder Minimum/LIA. Simulations with coupled atmosphere-ocean GCMs (global climate models; see Chapter 3) do, however, generally show a response in which the strength of the NAO signal correlates positively with solar irradiance.

Analyses of instrumental records of surface temperature and pressure on the 11-year solar cycle timescale show regional variations in the solar signal consistent with those found on longer timescales. Atmospheric blocking events, during which the jet stream is diverted in a quasi-stationary pattern associated with cold winters in western Europe (consistent with a negative NAO scenario), occur more frequently when the Sun is less active.[49] Allowing for a few years' lag between solar forcing and atmospheric response appears to strengthen this relationship.[50]

A robust solar signal is also seen in the Pacific Ocean, where a large positive anomaly in sea-level pressure in the northeast Pacific and a negative anomaly at lower latitudes indicates that the Aleutian low-pressure region sits farther to the west and the Hawaiian high farther to the north, and pushing the storm tracks further north in response to a more active Sun.[51,52]

In the Southern Hemisphere analyses of surface pressure measurements at mid–high latitudes, as well as model simulations, suggest that the Southern Annular Mode (SAM; see Chapter 3) is more often in a positive phase when the Sun is more active, with colder polar temperatures and stronger circumpolar winds. The picture is complicated, however, by an indication that the state of the Quasi-Biennial Oscillation (QBO; see Chapter 3) plays a role in the SAM response to the Sun.[53]

The polar response to solar variability is further discussed in Chapter 7, where it is suggested that these surface effects may be part of a response produced throughout the depth of the atmosphere but initiated by solar heating of the stratosphere.

Tropics

The signal in the North Pacific, noted in the previous section, is similar to the behavior associated in midlatitudes with a cold ENSO event (i.e., La Niña). A link to the solar cycle has been found in sea surface temperatures (SSTs) in the eastern tropical Pacific which is expressed as a cool (La Niña–like) anomaly at sunspot maximum, followed a year or two later by a warm anomaly, although the technique used to derive this result—namely, solar peak year compositing—means that there are only 14 data points, so the robustness of this signal has still to be established.

Analyses of tropical circulations are not conclusive, but a picture is emerging of a slight weakening and expansion of the Hadley cells[54,55] (see Chapter 2), so the descending branches, and dry air, occur at higher latitudes. There are also signs in precipitation indicators of a strengthening of the Walker circulation and strengthening of the Asian monsoon when the Sun is more active.[56] Consistent with these findings but on longer timescales, paleodata of precipitation derived from stalagmites in caves in southern Oman show lower monsoon precipitation associated with greater ^{14}C production (i.e., lower solar activity). On the solar cycle timescale, data reveal changes in tropical circulation, in cloudiness, and in the location and strength of regions of precipitation in the Pacific consistent with a cold ENSO-like signal at higher solar activity.

As outlined in Chapter 4, about 50% of the solar energy incident at the top of the atmosphere, averaged over the globe, reaches the Earth's surface but is not uniformly distributed. Radiation is most intense in the tropics, but most of it reaches the surface in the cloud-free subtropical regions. Over the oceans a large proportion of this radiant energy is used in evaporation. The resulting high-humidity air is carried by the prevailing trade winds into the tropics, where it converges with the stream from the other hemisphere and ascends, producing the deep cloud and heavy precipitation associated with the ITCZ. One mechanism for solar–climate links suggests that changes in the absorption of radiation in the clear-sky regions provide the driver: greater irradiance would result in enhanced evaporation, moisture convergence, and precipitation. These changes would result in stronger Hadley and Walker circulations and stronger trade winds, which would create greater upwelling in the eastern tropical Pacific Ocean and colder SSTs, and thus the observed La Niña–like signal described earlier.[57] It is not obvious that the amounts of energy involved in TSI variations (of order tenths of one percent) are sufficient to produce the desired effect, but there is some evidence for its taking place in GCM simulations of paleoclimate, and also over the solar cycle, although the timing of the signal relative to the cycle peak remains contentious.[58] It is also uncertain whether variations in the temperature profile within the lower atmosphere are consistent with what might be anticipated through this mechanism. Nevertheless, it remains a plausible candidate for a solar influence on regional climate and will benefit from further investigation.

Having considered the effects of solar variability at the Earth's surface, we now look higher into the atmosphere.

7 SOLAR INFLUENCE THROUGH THE ATMOSPHERE

SOLAR SIGNALS DETECTED IN OBSERVATIONS

Wind and Temperature in the Upper Atmosphere

At very high altitudes (higher than approximately 200 km) variations in temperature can be estimated from measurements of density deduced from atmospheric drag on spacecraft. In this very rarefied region the large variations in short wavelength (EUV) radiation from the Sun produce solar cycle changes in temperature of more than 400 K. One interesting observation from the density measurements, which have been made for more than 40 years, is a sharp drop (of about 28%) between that measured at the solar cycle minimum of 1996 and during the subsequent minimum in 2008/9.[59] The reasons for this drop probably include the long-term decline in solar activity noted in Chapter 5, as well as increases in atmospheric CO_2 (which cools the atmosphere at high altitudes), but other factors—as yet undetermined—may also play a role.

At lower heights two factors combine to reduce the amplitude of the solar cycle response in temperature:

first, the radiation reaching any particular level is attenuated by the atmosphere above, and second, the longer wavelengths that reach farther down are modulated less by solar activity. Thus at around 100 km altitude solar cycle variability is around 3–10 K, with a strong seasonal and latitudinal dependence. Measurements over the Antarctic show that these variations lead to a solar cycle amplitude in meridional wind of approximately 50% of its mean value.

Wind and Temperature in the Stratosphere and Troposphere

In pioneering work carried out by Karin Labitzke in Berlin during 1980s, measurements made over several decades from meteorological balloons across the globe were used to study variations in the average temperature of the atmosphere up to the lower stratosphere. Correlations between this temperature and solar activity, usually indicated by the solar 10.7 cm radio wave flux, were estimated, and a strong relationship was identified in mid-latitudes, implying temperature differences in that region of up to 1 K between minimum and maximum of the 11-year solar cycle.[60] This response was very intriguing, as it was larger than would be expected based on an understanding of variations in irradiance alone. Continued observations, including from space-borne instruments, confirmed this signal over subsequent solar cycles.[61]

Subsequent work in this area sought to isolate the solar effect from other possible influencing factors using multiple linear regression analysis. An example is given in Figure 7.1, which presents the solar cycle signal derived from a 40-year data set of temperatures throughout the troposphere and stratosphere, averaged around latitude circles. This is a "reanalysis" data set; it is based mainly on atmospheric measurements, but where observations are more sparse (especially at higher altitudes) the results of climate models are used to fill the gaps. Care is needed in interpreting the signals derived in these regions, as they may say as much about what was assumed for the model experiments as what was present in the atmopshere. The effects of other factors (a linear trend for climate change, QBO, ENSO, and stratospheric aerosol) were simultaneously extracted and excluded. In the tropics the solar signal in the stratosphere shows a structure with a double peak in the vertical profile: warming of up to 1.5 K at altitudes about 38–50 km; a smaller signal, less than 0.25 K, near 30 km; then slightly larger values in the lower stratosphere, especially in the subtropics. Other analyses[62] have not found the secondary peak in the lower stratosphere and conclude that the apparent lower solar peak is an artifact of the multiple regression technique employed and/or aliasing with the effects of volcanic eruptions.[63]

From the lower stratosphere, lobes of warming extend down into the midlatitude troposphere, with a small negative signal at lower latitudes. This pattern has also been detected in other data sets using multiple linear regression,[64] but an alternative analysis using empirical mode decomposition[65] shows warming throughout the

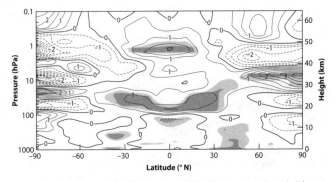

Figure 7.1 Difference in annual mean zonal mean temperature (K) between solar cycle maximum and minimum obtained from multiple linear regression analysis of European Centre for Medium Range Weather Forecasts ERA-40 reanalysis data set. Shaded regions are deemed statistically significant at the 5% and 1% levels. (Frame and Gray, 2010)[66]

troposphere. The former pattern is likely to be associated with an effect forced from higher levels (see the following discussion), whereas the latter pattern might be associated with heating through contact with a warmer sea surface. Thus the identification of the reasons for this discrepancy should contribute to an understanding of the mechanisms of solar–climate links as well as the suitability of different analytical techniques.

A similar analysis of west-east zonal winds shows that when the Sun is more active, there is a response in the winter upper stratosphere: an intensification of the strong circumpolar winds propagates downward. In the troposphere the midlatitude westerly jets are positioned slightly farther poleward when the Sun is more active, as shown in Figure 7.2. Panel (a) presents the observed

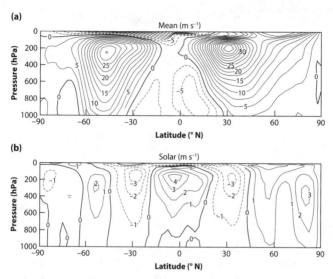

Figure 7.2 (a) Zonal mean zonal wind (m s⁻¹) for Dec–Feb as a function of latitude and pressure altitude from a National Centres for Environmental Prediction (NCEP) reanalysis climate data set. (b) Difference between solar maximum and minimum of the 11-year cycle derived using multiple linear regression. (Haigh and Blackburn, 2006)[67]

mean zonal wind (note that the vertical axis is pressure) with the strong jets in midlatitudes near the tropopause. Panel (b) shows the difference between solar maximum and minimum over the 11-year solar cycle: the jets have weakened on their equatorial side and strengthened on the polar side, exhibiting a poleward shift in both hemispheres.

These observations are consistent with the regional signals at the surface discussed in Chapter 6: at higher solar activity the tropical overturning Hadley cells spread out, moving the subtropical high-pressure regions and midlatitude storm tracks a little toward the poles. These responses cannot be explained simply by changes in solar radiation; a possible causal link is outlined later in this chapter.

Stratospheric Ozone

We are interested in knowing how the Sun affects ozone for a number of reasons. One key concern relates to understanding how legal limitations on the emissions of ozone-depleting substances, such as chlorofluorocarbons, have been able to constrain polar ozone depletion (the "ozone hole"). To assess the impact of the legislation necessarily involves understanding how natural factors influence ozone, in a way analogous with the detection of human influences on global warming. Ozone is also fundamental to climate in determining both the temperature of the stratosphere and the flux of radiation to lower levels, so it plays a role in radiative forcing.[68] In addition, its heating effects may play an indirect part in determining how the Sun influences the lower atmosphere, through dynamic coupling between different atmospheric layers; this idea is developed later in this chapter.

There is plenty of evidence that stratospheric ozone responds to solar activity on the 11-year cycle timescale. Analyses of ozone measurements from ground- and space-based instruments suggest that the amount of ozone within a vertical column of atmosphere varies by 1%–3% in phase with the solar cycle, with the largest signal in the subtropics.[69] The vertical distribution

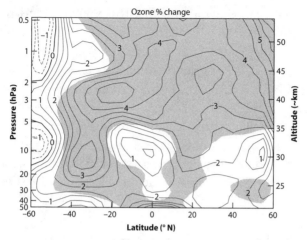

Figure 7.3 Annual mean solar cycle regression coefficient (change from solar minimum to maximum) for Stratospheric Aerosol and Gas Experiment (SAGE II) ozone profile data over the period December 1984–August 2003. Shaded areas are significant at the 95% confidence level. (Hood and Soukharev, 2012)[70]

of the solar signal in ozone is more difficult to establish, because of the short length of individual observational records and problems with intercalibration of the various instruments. In the tropics, analysis of observations from satellites suggests a signal in phase with solar activity such that in the upper stratosphere ozone concentrations are 2%–4% higher at solar cycle maximum relative to the minimum, with a 1%–2% signal in the lower stratosphere, and a smaller response in between, as illustrated in Figure 7.3. This response is associated with the double-peak temperature profile discussed earlier and is similarly awaiting a full interpretation.

The rotation of the Sun (see Chapter 3) modulates the radiation reaching the Earth because of the distribution of sunspots and faculae on its surface. With a periodicity of about 27 days the length of the data record needed to establish these signals is short, compared with that required to detect 11-year cycle variability, and signals of solar rotation as well as temperature have been detected in stratospheric ozone.[71] The difference in the amplitude of the 27-day signal between periods of greater and lesser activity over the 11-year cycle provides a useful test for models of stratospheric chemistry and climate.

UNDERSTANDING THE SIGNALS

We have seen some of the circumstantial evidence for an effect of solar variability on the atmosphere and climate. We now consider how changes in the Sun influence the temperature and composition of the stratosphere and how they might lead to a "top-down" effect producing some of the meteorological effects outlined in the preceding section.

Variations in the Solar Spectrum

We start by looking in more detail at how solar radiation is transmitted and absorbed as it passes through the atmosphere. Transmission and absorption are determined by the spectral properties of the component gases and so are strong functions of wavelength. Figures 3.2 (a) and 4.4 (c) show that very little of the incident radiation

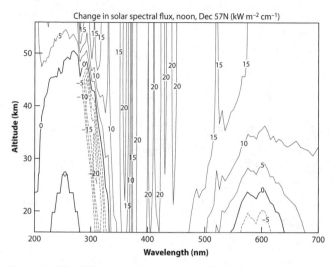

Figure 7.4 The difference between 2007 and 2000 (representing the minimum and maximum, respectively, of the last solar activity cycle) in solar spectral irradiance (kW m^{-2} cm^{-1}) in the ultraviolet and visible regions as a function of wavelength and altitude in the atmosphere, calculated for a clear sky at latitude 57° N on 21 December at noon.

at wavelengths shorter than about 300 nm reaches the ground. They also indicate wavelength regions in the infrared region where solar radiation is absorbed, mainly by water vapor. Figure 4.5 indicates how the absorption of ultraviolet and visible radiation translates into stratospheric heating rates.

Figure 7.4 shows the field of spectral irradiance for the difference between solar cycle maximum and minimum conditions, based on the spectral variability of Figure 3.2 (b). In these plots (which were produced using a model of atmospheric circulations and photochemistry) the effects of changes in ozone concentration resulting from the enhanced solar irradiance are included. This means that the vertical penetration of the enhanced irradiance is not spectrally uniform, because it depends on the perturbation in ozone concentration as well as its absorption spectrum. Much of the increased irradiance penetrates to the surface but at wavelengths less than 320 nm the increases in ozone result in lower values of irradiance throughout the stratosphere, despite the increased insolation above. In similar fashion, at wavelengths in the 550–640 nm range, less radiation reaches the lower stratosphere. This effect is more marked when the Sun is lower in the sky and so leads to strong latitudinal gradients in the spectrally integrated irradiance in winter midlatitudes. These results give an indication of the complexity in solar heating rates introduced by variations in solar variability due to details of the photochemical response in ozone.

Coupled chemistry-climate models provide important tools for investigating the processes involved in a solar influence on the stratosphere. A collaborative study[72] compared estimates from different models of the response to solar variability in heating rates, temperature, and ozone fields and concluded that models that do not account for variations in the spectrum of solar irradiance cannot properly simulate solar-induced variations in stratospheric temperature. Nevertheless, even the models which do incorporate more realistic representations of the solar spectrum produce mixed results in terms of reproducing the apparent observed profiles. They also employ a variety of assumptions concerning factors such as sea surface temperatures, the representation of the QBO, and vertical resolution, making interpretation more complex.

Data from the SORCE satellite suggest that the solar cycle variation in UV may be larger than previously assumed[73] (and were used in the preceding calculations), though the data are currently in the process of being recalibrated (see Chapter 5). Model studies indicate that they will imply significantly different effects on stratospheric composition.[74] Given the key importance of space-based measurements in determining the response to solar variability of the middle atmosphere, as well as possibly the climate at lower levels (see further discussion), it is of concern that they may be interrupted for several years owing to a lack of available missions.

Solar Energetic Particles

On shorter timescales ozone is also affected by the energetic particles emitted during solar storms, flares, and coronal mass ejections, as outlined in Chapters 5 and 8. Intense *solar proton events* (SPEs) produce high-energy particles which follow the Earth's magnetic field lines toward the poles into the upper atmosphere, where they ionize and dissociate nitrogen and water molecules, producing NO_x (i.e., N, NO, and NO_2)[75] and HO_x (i.e., H, OH, and HO_2).[76] These species destroy ozone through the catalytic cycles outlined in Chapter 4. Measurements from satellites show large enhancements in the concentration of NO_x following intense SPEs, as illustrated in the upper panel of Figure 7.5, which presents the production of NO_y (total reactive nitrogen, i.e., NO_x plus the compounds produced by its oxidation, including NO_3, HNO_3, and N_2O_5) by SPEs over a 10-year period, calculated in a GCM based on proton flux data from satellites. More events occurred around the peak of the solar cycle in 2000/2001, but they were clearly very intermittent. The air enhanced in NO_y propagates downward over a period of months into the polar lower stratosphere, where it can lead to decreases in polar ozone. Model calculations of the fractional changes in NO_y and O_3 are shown in the lower panel of Figure 7.5. The possibility that solar energetic particle events are able to influence the climate of the lower atmosphere on decadal timescales is the subject of current research.

Galactic Cosmic Rays and Clouds

It has been proposed that galactic cosmic rays (GCRs) could affect solar influence on climate through their role in atmospheric ionization. Consequently, it has been further suggested that long-term trends in the incidence of cosmic rays might be responsible for "global warming." There is no robust experimental evidence for a solar signal in large-area cloud cover, but it is instructive to consider how GCRs might have an effect.

GCRs produce ionization throughout the lower atmosphere, and it is well documented that this process is

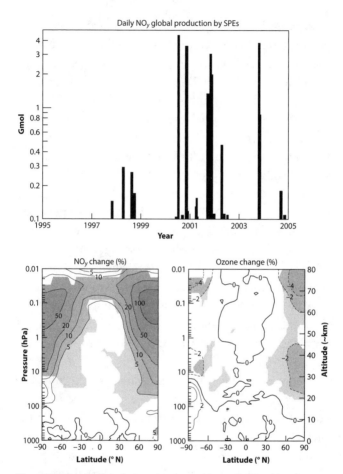

Figure 7.5 Top: Daily column nitrogen oxide production by solar proton events (SPEs) in gigamoles (6.02×10^{32} molecules) as a function of time for calendar years 1995 through 2004. Bottom: The effects of SPEs averaged over 2000–2004, relative to years with no SPEs on (left) nitrogen oxides (contour intervals 0, 5%, 10%, 50%, 100%) and (right) ozone (−5%, −2%, 0, 2%, 5%); shaded regions indicate 95% statistical significance with Student's t-test (Jackman et al., 2009)[77]

modulated by solar activity over the 11-year cycle—with greater ionization when the Sun is less active because of the inverse relationship of cosmic rays with solar activity (as outlined in Chapter 3). Two different routes might result in this ionization influencing cloud cover.[78] The first involves the preferential growth of ionized particles to a size which is energetically favorable for cloud droplet formation (see Chapter 2). Subsequent to the ionization, several consecutive processes need to occur to result in the necessary enhancement in concentration of cloud condensation nuclei. A specially designed experiment at CERN[79] has provided suggestive evidence for ion-induced nucleation or ion-ion recombination as sources of aerosol particles. Limits on the ability of the experiment to create realistic atmospheric conditions, however, mean that it is not possible to conclude whether this mechanism could produce a significant impact on cloud cover in the context of other droplet-growth mechanisms.

The second path whereby ionization might influence cloud cover involves changes in the global electrical circuit. This circuit involves currents flowing between the surface and the ionosphere initiated by thunderstorm activity and completing the circuit in fair-weather regions. The presence of (nonthunderstorm) cloud in the clear regions increases atmospheric resistivity, and there is some evidence of modulation due to ionization by cosmic rays. Charge can accumulate near the edges of clouds, which can influence both evaporation of cloud droplets and interactions between them. Again, the mechanisms are plausible, but evidence for a significant impact remains elusive.

Decade- to century-scale trends in GCRs do not correlate well with global mean surface temperature, and this fact, together with the very tenuous evidence for any relevant mechanisms, leads us to conclude that ionization by cosmic rays is unlikely to play a significant role in climate change.

Dynamic Effects within the Stratosphere

It has been realized since the mid-1970s that variations in solar heating would influence the thermal structure of the middle atmosphere. Early attempts to predict the response were carried out using two-dimensional (latitude–height) chemistry transport models, such as used to estimate the solar radiation field in Figure 7.4. These models predicted peak warming near the stratopause, and largest increases in ozone at altitudes around 40 km, with perturbations in both fields monotonically decreasing toward the tropopause. They did not reproduce the more complex latitudinal and vertical gradients shown in Figures 7.1 and 7.2 (b). If the structure shown in these figures represents the true solar response, then it must be mediated through changes in atmospheric circulation as well as through direct radiative impacts. Furthermore, as the structure in the temperature signal is fundamentally related to the ozone response, it is unlikely that any simulation would satisfactorily reproduce the one without the other.

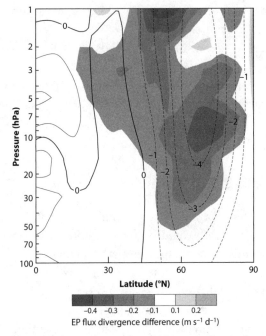

Figure 7.6 Modeled response to solar UV forcing in the Northern Hemisphere for January–February. Difference between solar minimum and solar maximum. Contours: zonal mean zonal wind (m s⁻¹); shading: acceleration of zonal flow by large-scale waves (Ineson et al., 2011)[80]

An increase in the temperature difference between equatorial and polar regions, introduced by enhanced solar heating near the stratopause, tends to strengthen the westerly winds blowing in the winter hemisphere. Conversely at low solar activity (Figure 7.6) an easterly anomaly moves downward and toward the pole through the winter, extending into the lower atmosphere in spring.

Such solar-induced changes in the zonal wind structure affect the upward propagation of planetary waves (see Chapter 2), and it is likely that the interaction of these with the background wind structure produces the downward-moving feature. The momentum deposited by the planetary waves also influences the strength of the mean overturning of the stratosphere and so further affects the temperature and wind structure (see schematic of Figure 7.8). With greater solar heating near the tropical stratopause the stratospheric jets are stronger, the polar vortexes less disturbed, and the overturning circulation weaker. This mechanism produces cooling in the polar lower stratosphere owing to weaker descent, and warming at low latitudes through weaker ascent. Thus in the lower stratosphere the tropics are warmer than they would be from radiative processes alone, and the poles are cooler, further strengthening the circumpolar winds, as described earlier.[81]

Models including fully interactive chemistry, so that the imposed irradiance variations affect both the radiative heating and the ozone photolysis rates, allow feedback between ozone concentrations, temperature, and transport. These models simulate an improved vertical structure of the annual mean ozone signal in the tropics. Some models include the lower stratospheric maximum, although no clear picture has emerged identifying the factor responsible for the apparent improvement in structure. Candidates include time-varying sea surface temperatures, transient solar input, ability of the model to produce an internally generated QBO, high vertical resolution, and inadvertent aliasing with the signal of ENSO.

A Dynamic Influence of the Stratosphere on the Troposphere

As noted in Chapters 3 and 5, solar radiation at UV wavelengths varies by a larger fractional amount than does visible radiation (or TSI). UV radiation is largely absorbed in the atmosphere above the tropopause, so that any influence it has on climate must be through a top-down mechanism. This process has been tested by experiments with climate models such as illustrated in Figure 7.7, which shows the response of storm tracks in a GCM to changes in UV radiation over the 11-year cycle as seen in Figure 3.2 (b). In this model experiment sea surface temperatures were fixed, eliminating the possibility of a bottom-up effect. The storm tracks shift poleward at solar maximum, consistent with the shifts in the jets in observational data (Figure 7.2), so the model is—at least qualitatively—successful in simulating the tropospheric patterns of response to solar variability. The model signal is weaker than observed but is intensified when larger increases in stratospheric ozone are imposed. The model study clearly reveals a dynamic influence of changes in the stratosphere on the troposphere rather than a direct radiative effect.

More generally, a number of different studies have indicated that such a downward influence does exist. Observational analyses suggest a downward propagation of polar circulation anomalies in the Northern and Southern Hemispheres. Models have also demonstrated a downward influence of Antarctic stratospheric ozone depletion on the circumpolar circulation in the Southern Hemisphere, and

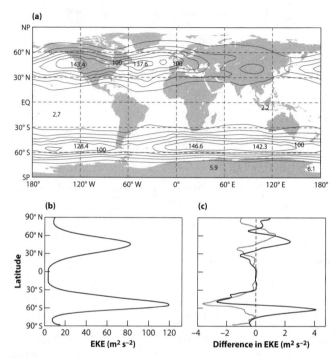

Figure 7.7 (a) Storm tracks in a climate model, indicated by transient eddy kinetic energy (EKE) ($m^2 s^{-2}$) near the tropopause. (b) Average of (a) around lines of latitude. (c) As (b) but the difference between solar maximum and minimum specified in the model through a change in ultraviolet radiation, with (black curve) and without (gray curve) an imposed change in ozone. (Haigh, 1996)[82]

of stratospheric temperature trends on the NAO. These studies did not specifically address the impact of solar variability on climate, but they did suggest that the troposphere responds to perturbations initiated in the stratosphere. The study introduced in Figure 7.6[83] using a coupled atmosphere–ocean GCM specified large changes (as suggested by early versions of data from the SORCE satellite) to solar radiation in the 200–320 nm range and produced both the downward-moving winter wind anomaly outlined earlier and, at lower solar activity, a significant shift to a negative NAO pattern and colder winters in western Europe, similar to signals found in observational records.

Many mechanisms have been proposed whereby the lower stratosphere may exert a dynamic influence on the troposphere.[84] These processes include a response of the mean overturning circulation to angular momentum forcing from above, modification of the refraction or reflection of upward-propagating planetary-scale waves, and feedbacks between changes in the mean winds and tropospheric weather systems. Coupling between the Hadley circulation and midlatitude waves may also play a key part: a number of modeling studies have indicated a link between broadening of the Hadley cells and poleward shift of the storm track, and additional heating introduced by solar-induced ozone into the tropical upper troposphere and lower stratosphere.

MECHANISMS FOR A SOLAR INFLUENCE ON CLIMATE: AN OVERVIEW

In Chapters 6 and 7 we have seen how signals of solar variability are expressed in the climate system and considered how the apparent effects are produced. Here we present an overview of the potential mechanisms, which are also summarized by the schematic in Figure 7.8.

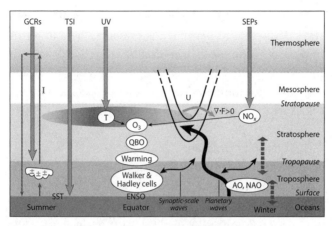

Figure 7.8 A schematic indicating many of the mechanisms proposed for a solar influence on climate and discussed in the text. The layers represent different regions of the atmosphere and oceans. Most solar radiation (TSI) passes through to the surface, where it warms the land and raises sea surface temperature (SST), potentially influencing the overturning circulations of the tropical lower atmosphere (the Walker and Hadley cells) and interacting with ENSO. Solar UV radiation is mostly absorbed in the stratosphere, where it increases temperature (T) and produces ozone (O_3). The resulting change in latitudinal temperature gradient influences the upward propagation of large-scale atmospheric waves (thick black wiggly line), affecting their deposition of momentum ($\nabla \cdot F$), the strength of the wind (U) in the polar vortex, the overturning circulation of the middle atmosphere, and temperatures in the tropical lower stratosphere, which may influence the quasi-biennial oscillation (QBO). Stratosphere–troposphere interactions may also be mediated through synoptic-scale waves associated with midlatitude weather systems (thin black wiggly line) and may affect polar modes of variability (AO, NAO). Energetic particles from solar storms and flares produce nitrogen oxides (NO_x) in the polar middle atmosphere, which affect ozone concentrations and possibly temperature and circulation. Galactic cosmic rays, which are modulated by solar activity, ionize the atmosphere and affect the global electric circuit. They may also influence cloud cover by enhancing the growth of condensation nuclei. (Gray et al., 2010)[85]

Three aspects of solar variability are involved, with several different routes. Each is presented here in terms of climate response to increased solar activity, with an inverse response to reduced activity implied. These processes are not mutually exclusive, and two or more processes may be operating simultaneously.[86]

1. Total solar irradiance
- **Earth energy balance**: An increase in total solar irradiance results in an enhancement of the energy entering the climate system and a global average warming at the Earth's surface. At equilibrium a 1 W m^{-2} increase in TSI will produce an increase in global average surface temperature of approximately 0.1 K. This mechanism is essentially indisputable but gives no indication of the emergence of regional responses.
- **Sea surface temperatures**: Enhanced solar heating of the sea surface in the tropical Pacific enhances evaporation and precipitation, and strengthens the Walker cell. The increased wind stress at the ocean surface results in greater upwelling of cold ocean water in the East Pacific. The response, which resembles a La Niña event, lags by a few years. As in ENSO events, there is an associated change in the meridional overturning of the upper ocean and its associated impacts across the Pacific. This mechanism accounts for some observational records but is premised on very small fractional variations in

TSI and awaits confirmation of details of the processes involved.

2. Impacts on the middle atmosphere

- **Ultraviolet radiation and the stratospheric polar vortex**: Enhanced solar UV radiation warms the middle atmosphere, particularly near the stratopause in the tropics and summer subtropics. The enhanced meridional temperature gradient strengthens the winter stratospheric polar vortex. Through interaction with upward-propagating planetary waves, this signal in wind can propagate down to the surface, appearing, for example, as a positive phase of the North Atlantic Oscillation.

- **Solar energetic particles and middle atmosphere composition**: Energetic particles emitted during solar flares or coronal mass ejections arrive at Earth and are steered to the poles by the geomagnetic field. Their impact with atmospheric molecules results in the production of, particularly, nitrogen oxides, which enter into ozone-destroying chemical reactions. The reduction in polar ozone may produce changes in temperature gradient and the polar vortex in the same way as described in the previous paragraph. However, the effects are intermittent and less well documented than those observed in response to UV.

- **Ultraviolet radiation and the tropical lower stratosphere**: UV radiative heating of the middle atmosphere also weakens the meridional overturning circulation there, which further warms the lower stratosphere in the tropics. The resulting changes in thermal structure around the tropopause result in a weakening and expansion of the tropical tropospheric Hadley cell and a poleward shift of the midlatitude jet streams.

- Although details of these mechanisms are still not fully established, it is becoming clear from other (nonsolar) studies that the stratosphere can influence the troposphere through dynamic coupling and that this pathway is feasible for solar effects.

3. Ionization by galactic cosmic rays

- **Ion-induced growth of cloud condensation nuclei**: Cloud droplets form when water vapor condenses onto cloud condensation nuclei (CCN). CCNs of sufficient size are created by the condensation of sulfuric acid and organic vapors onto very small aerosol particles, and this process is enhanced by the presence of ions. Thus when the Sun is more active, and ionization by cosmic rays is reduced, there may be fewer CCNs, making clouds less reflective, reducing Earth albedo, and warming climate.

- **Global electric circuit and near-cloud aerosol ionization**: The clear-sky electric current flowing between the ionosphere and the ground is interrupted by clouds, causing a buildup of charge—positive at the cloud top and negative at the base. Aerosols near the cloud boundaries also become highly charged, migrate within the clouds, and possibly enhance the formation of ice particles.

When the Sun is more active, a reduction in atmospheric ionization by GCRs reduces the electric current and may affect ice particle formation.

More evidence is required before either of the proposed pathways can be considered both operative and sufficiently effective to affect climate.

8 SPACE WEATHER

IN CHAPTER 3 WE NOTED THE OCCURRENCE IN THE solar corona of energetic phenomenon such as flares and coronal mass ejections (CMEs) which can have a major impact on the Earth's space environment. This link between Sun and Earth is generically known as *space weather* and has a very old history. For example, auroras have been known for millennia, and on September 1, 1859, a causal effect between a large solar flare and a magnetic disturbance at the Earth was identified by Carrington. There were other discoveries in subsequent years, but the 1950s and 1960s brought major advances in the understanding of the connection between the Sun and the Earth. Satellite observations confirmed the existence of the solar wind, originally proposed theoretically by Parker in 1958 (see Chapter 3), so that the nature of the interplanetary medium was identified and measured. The Earth's magnetic field forms a cavity in the solar wind (the magnetosphere). The proposal of Dungey[87] that the magnetic fields of the Earth and solar wind could interact, allowing solar wind mass and energy into the Earth's protective magnetosphere, was crucial in understanding the effect of the Sun and solar wind on the terrestrial space environment, though it took a surprisingly long time to be accepted generally.

A vast number of observations of (a) the Sun by ground- and space-based instrumentation, (b) the solar wind by in situ spacecraft, (c) the magnetosphere by orbiting spacecraft, and (d) terrestrial phenomena such as auroras led, by the mid-1990s, to a clear picture of the link between the Sun and space weather. Continuous near-real-time measurements of the Sun and solar wind from the upstream L_1 (Lagrange) point (roughly 1.5×10^6 km sunward of the Earth) have been made by the Solar and Heliospheric Observatory (SOHO) and WIND spacecraft since 1995 and 1994, respectively, with the latter superseded by the Advanced Composition Explorer (ACE) spacecraft in 1997. Such continuous monitoring of the Sun and solar wind has, in turn, led to methods for predicting deleterious space weather.

CAUSES OF SPACE WEATHER

Space weather is caused by three agents: (a) fluctuations in the solar wind magnetic field and velocity; (b) energetic charged particles which may be of solar or galactic origin or intrinsic to the magnetosphere; and (c) solar radiation, especially UV. The major solar causes of space weather are CMEs and associated solar flares (see Chapter 3). The shock driven ahead of a fast CME accelerates charged particles (mostly protons, referred to as *solar energetic particles*, or SEPs), and these can reach the Earth within minutes of CME onset. The CME itself arrives 2–4 days after onset, and the key to its impact (or *geoeffectiveness*) is the orientation of its magnetic field and its velocity. When the CME magnetic field (or interplanetary magnetic field; IMF) is southward (i.e., pointing toward the south ecliptic pole), it interacts with the Earth's field, as proposed by Dungey and, as we discuss shortly, permits direct access of energy to the magnetosphere. When the field is northward, such an interaction does not occur. Similar, though less drastic, processes occur when fast and slow regions of the solar wind interact, compressing any magnetic field caught up in the interaction. (It should also be noted that galactic cosmic rays (GCRs; discussed in Chapter 5) are a major nonsolar space weather hazard, though their level is determined by the level of solar activity.)

In the absence of the solar wind, the Earth's magnetic field would be roughly dipolar (with the exception of aspects such as the South Atlantic Anomaly). The interaction with the wind leads to distortion of the field into a magnetosphere, as shown in Figure 8.1 (a). The magnetosphere is the major shield from the electromagnetic field and charged particles of space weather. Planets with only residual fields, such as Mars, have undergone atmospheric erosion from the wind interaction with an unprotected atmosphere.

The magnetosphere extends roughly 10 R_E on the sunward side and has a magnetic field strength varying from of order 5×10^{-5} T at the Earth's surface to 20 nT or so at the boundary with the solar wind (the magnetopause). A *magnetotail* extends over 100 R_E in the direction away from the Sun. For northward IMF, the solar and terrestrial fields do not interact on the day side but do so at the

(a)

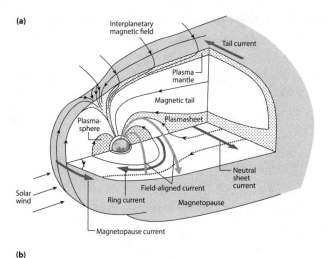

(b)

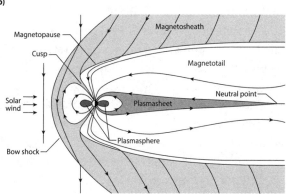

Figure 8.1 (a) A three-dimensional visualization of the magnetosphere, with the important current systems highlighted. The magnetopause is roughly 10 R_E on the sunward (left) side, and the magnetotail extends tens of Earth radii on the night side (right). (b) A two-dimensional cut through the noon–midnight plane showing the interaction between solar and terrestrial magnetic fields through the magnetic reconnection process at the magnetopause on the left, and in the tail on the right.

high-latitude cusps. For southward IMF, a process known as *magnetic reconnection* permits the magnetic field to change topology: solar wind magnetic field lines become connected to terrestrial ones, and vice versa (Fig. 8.1 (b)). The interconnected field lines are swept downstream by the solar wind and are dragged out to form the long magnetotail. In the tail, the field lines north and south of the magnetic equator are now oppositely directed, and a second reconnection process occurs, perhaps 30 R_E downstream of the Earth. The field connectivity changes again, and magnetic flux is swept away into interplanetary space but is also convected toward the Earth. The latter moves around the Earth, returning to the day side, completing the cycle first predicted by Dungey.[88]

The overall energy flow in the magnetosphere is complicated and is summarized in Figure 8.1 (a), and is also described in various articles in Kamide and Chian.[89] A series of current systems form a closed global circuit. Of these, the most important for the present discussion are the ring currents (or *van Allen belts*), two (and sometimes three) regions of energetic charged particles located between 2 and 6 R_E, and the field-aligned currents that close through the ionosphere and below. Prolonged exposure to a strong southward IMF leads to strong enhancement of these systems and is termed a *major geomagnetic storm*. In the interplanetary medium, such storms begin shortly after flare/CME onset with the arrival of SEPs. The intensity of such particle streams can remain enhanced until the CME itself arrives. The geomagnetic storm in the magnetosphere starts abruptly as the CME and its leading shock hit the Earth's magnetic field. Strong currents flow to close the global circuit and, over a longer time, the ring currents build in intensity. Once the CME has passed, the ring current decay can take up to a day.

What scale of geomagnetic storm should cause worry? At the maximum of the solar cycle, CMEs are frequent, but major space weather–related problems are not so common (excluding classified military concerns). However, recent near misses and realization of the harm a major storm could cause suggest that efforts should be focused on the very rare massive geomagnetic storms (superstorms; e.g., Cannon et al.[90]). The best known of these remains the disturbance associated with Carrington's flare on September 1, 1859, but a major event also occurred in February 1956, and near misses occur from time to time. The concern is that although such storms were a curiosity in the pre–space age past, their consequences today would be very serious given the technological aspect of modern life. This thinking represents an important step away from the desire to forecast precise conditions—as seem to be required in many military applications—to concentrating on those events that might cause widespread disruption or worse to a wide range of technical systems.

EFFECT OF SPACE WEATHER ON TECHNICAL SYSTEMS

Figure 8.2 shows various systems that are subject to disruption by space weather. The important ones can be summarized as follows.

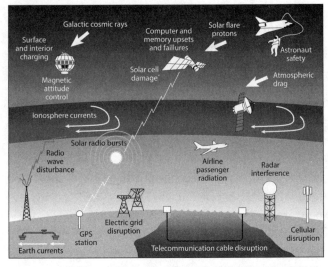

Figure 8.2 A schematic of space weather effects that are described in the text. The vertical distances are not to scale but show near-Earth (ground and lower atmosphere), ionosphere, and inner magnetosphere. (Adapted from an image courtesy of NASA Goddard Space Flight Center)

Geomagnetically Induced Currents

During major geomagnetic storms, changes in the Earth's magnetic field at the surface induce currents. These currents can flow along good conductors such as power lines and cables, though the magnitude of this effect depends on many aspects, including the nature of the ground. This current can lead to transmission line problems as well as transformer failure. The famous Hydro-Québec incident of 1989, which led to a province-wide power failure, is the best example, but numerous other incidents have been documented. The timescale is of order minutes to hours.

Atmospheric Drag

Spacecraft in low Earth orbit are subject to a continual slow loss in altitude owing to a drag force exerted by the very upper reaches of the atmosphere and normally this is easily accounted for. However, drag can increase during a particularly active solar cycle owing to enhanced UV emission, shortening the satellite's life. On short timescales, rapid heating during geomagnetic storms can lead to an abrupt expansion of the outer atmosphere and a sudden change in satellite position, leading to tracking problems, which are well documented.

Airline Radiation Levels and Astronaut Safety

Airline and astronaut safety are considered together, since the problems are similar. Two issues are of concern. One is the radiation exposure due to solar energetic particles (SEPs), which typically occur during geomagnetic storms. The second is the cumulative exposure to galactic cosmic rays (GCRs), of which there tends to be a larger flux nearer solar minimum. Airlines can quantify the effects and take mitigating actions such as restricting the time crews are exposed or altering flight paths. The problem for astronauts, especially on proposed missions to Mars, is more difficult, since cumulative doses must be estimated and appropriate radiation shielding provided.

Spacecraft Anomalies

Single cosmic rays (single-event upsets) can scramble parts of computer memory, and more prolonged exposure to radiation—such as can occur in a geomagnetic storm—can result in discharges of charged particles either on the surface of or inside a spacecraft. The latter is particularly an issue for geostationary satellites operating in the proximity of the outer radiation belt. Numerous examples of problems have been reported, including several cases of total loss of spacecraft. Extensive hardening of components and additional shielding can mitigate these risks.

Communications

Communication problems are related to spacecraft anomalies, given the reliance on satellites for transmission of signals, such as global positioning data. However, historically there have been problems with communications at low latitudes owing to ionospheric scintillations arising from heating of the ionosphere.

PREDICTION AND FORECASTING OF SPACE WEATHER

The key to forecasting space weather is to predict large solar eruptions, if possible, or alternatively, to immediately identify their onset. In addition, it is very desirable to be able to detect, in real time, the arrival of the

CME in near-Earth space. Until the advent of the SOHO, WIND, and ACE spacecraft in 1996 such complete forecasting was impossible on a continual basis. Before then, ground-based observations of the Sun could detect the magnetic complexity associated with CMEs and flares, as well as the initial radiation response in, for example, the Hα (hydrogen-alpha) emission line. Occasionally, an orbiting spacecraft such as the Japanese Yohkoh mission could see the flare onset in X-rays, though in low-Earth orbit such spacecraft had regular data gaps, for example, at the South Atlantic Anomaly.

The essential instrument for detecting CMEs at the Sun is a coronagraph that observes it continuously, so one based in space is required. (A *coronagraph* is a telescope in which an occulting disc blocks out light from the solar disk, allowing images of the outer corona either in emission lines (such as Fe) of ionized plasma or through Thomson scattering. Since the intensity of the former scales as the density squared and the latter as the density, Thomson scattering is the preferred technique in the outer corona.) Ground-based coronagraphs have a long history (since 1940s) but are limited by weather, darkness, and other factors, so it is impossible to build a reliable forecasting system based solely on them. Research coronagraphs on spacecraft have flown since the early 1970s, but it was the SOHO spacecraft located at the L$_1$ point that first made continuous observations possible. Since 2006 the SOHO coronagraphs have been complemented by those on the twin STEREO spacecraft, permitting a multipoint view of CMEs and thus making better estimates of velocity and direction possible. Prompt analysis of coronagraph data now permits identification of CMEs shortly after eruption and also gives an estimate of their direction of motion. Empirical, numerical, and physics-based models now permit predictions of the arrival time of a CME at the Earth and, in some cases, the orientation of its magnetic field. Errors are typically 10 hours or so.

In situ monitoring of the solar wind at the L_1 point provides real-time information of the magnetic field vector and the properties of electrons, protons, and other ions. In particular, the start of the SEPs shortly after CME onset can be recorded. Perhaps in situ L_1 data are most useful for postevent analysis, although efforts have been made to develop real-time tools to forecast space weather, in particular the incoming magnetic field strength and profile.

Large-scale numerical models are a third and increasingly important aspect of space weather forecasting. The largest of these use a network of linked models to examine the evolution of, for example, a CME at the Sun, its journey through the solar wind, its interaction with the Earth's magnetosphere, and its consequences for the space environment. Whether such a "kitchen sink" approach is useful for understanding the relevant physics or for making useful forecasts is unclear, but it does seem to represent the chosen way forward, at least in the United States.

Finally, how is a given forecast judged to be a success? Various metrics are being developed, in particular by the U.S. space weather community, for forecasting tools. In the interplanetary medium these include the arrival time, magnetic field strength, and orientation of CMEs, and the associated magnetospheric and terrestrial responses. As noted earlier, these requirements have led to very sophisticated modeling. The real answer to the question depends on user needs.

SPACE CLIMATE

Space climate is a newer topic and refers to long-term changes in the Sun's magnetic activity level over decades or centuries. Certainly, the Maunder and Dalton Minima, discussed in Chapter 5, suggest that solar magnetic activity can change considerably, and it is an interesting question whether the recent "great minimum" is a prelude to another such event. The answer is not without importance for the technical systems influenced by space weather. Diminished activity would imply lower UV emission, less flaring activity, and so forth, which would suggest less need to worry about atmospheric drag on low-Earth-orbiting spacecraft, and SEPs for geostationary ones. However, low activity will increase the flux of GCRs, making single-event upsets more common and also perhaps raising issues of radiation dose for airline passengers. This topic is only now becoming a serious area of research.

9 SUMMARY

FUNDAMENTAL TO THE WELL-BEING OF HUMANITY IS the Earth's climate, and any factor with the potential to affect that is obviously of concern. Thus an understandable interest in the body that provides the energy for all life on Earth has driven a long history of study of how changes in the Sun might influence the climate. The wealth of physical, chemical, and biological processes involved also makes the topic of intrinsic scientific fascination.

Observations of the Sun, alongside theoretical advances and developments in models, are helping to further understanding of its behavior. In particular, significant advances have been made in determining how different activity indicators relate to the physical processes involved in the evolution of the solar magnetic field, sunspots, and radiation over the 11-year cycle. Satellite-mounted instruments are now providing measurements of irradiance, allowing quantitative assessment of solar radiative forcing of global temperature change over these time periods. On longer timescales there are still uncertainties in underlying trends in solar irradiance, partly owing to problems of instrumental calibration and intercalibration; as a result, reconstructions of past values of solar irradiance are still prone to uncertainty. Nevertheless, the results of statistical studies of global temperature records concur with those from climate models that while increases in solar activity might have contributed up to 10% of the global warming that took place from the mid-nineteenth to the mid-twentieth century, the warming over the latter half of the twentieth century is due almost entirely to the increasing concentrations of greenhouse gases from human activity.

Solar activity affects the Earth in a number of other ways. The energetic particles emanating from large geomagnetic storms create changes in the magnetosphere which can induce currents in conductors on the surface of the Earth. The effects of this space weather have included damage to power supplies as well as to spacecraft operation and communications. Forecasting large solar eruptions is very difficult, but understanding how the subsequent effects might develop means that real-time monitoring of the solar corona can provide some warning of potential space weather impacts.

Solar signals have been documented in various regions of the atmosphere over certain parts of the globe on a range of timescales. In the upper atmosphere there are large variations in temperature and wind over the 11-year cycle. In the middle atmosphere stratospheric temperature and ozone demonstrate a clear positive correlation with solar irradiance and also, at high latitudes, a response to energetic particle events.

Signals of solar variability near the Earth's surface are smaller and more difficult to establish, but evidence is accumulating that changes in the Sun are associated with increased likelihood of certain circulation patterns. Around the peak of the 11-year cycle the overturning circulations in the tropics present greater latitudinal extent, and the midlatitude storm tracks move slightly toward the poles. Consistent with this pattern is evidence that during periods of lower solar activity the North Atlantic Oscillation is somewhat more likely to be in a negative phase, so that the jet stream, meandering across the North Atlantic Ocean, is positioned to create colder winters in western Europe. In the eastern tropical Pacific Ocean there are indications that higher solar activity is related to cooler sea surface temperatures in a pattern akin to that seen in negative phases of the El Niño–Southern Oscillation. On centennial to millennial timescales, other studies show a link between low solar activity and weaker winds in the southern Asia monsoon. Nearly all these apparent solar signals require further data or more robust statistical analysis to confirm their validity.

Understanding the physical processes involved in solar–climate connections is crucial to interpreting meteorological records and also to predicting aspects of the future climate. Several mechanisms have been proposed to explain the observed regional near-surface effects of solar variability, and scientific research is making advances in understanding each of these mechanisms. A theory involving heating of the stratosphere by solar UV radiation, with effects transmitted by changes in winds and circulation down to the surface, is the most well developed in explaining midlatitude signals. Another theory, involving solar heating of the cloud-free ocean surface and feedback on overturning circulations in the atmosphere and ocean, may provide a clue to tropical effects. These processes are not mutually exclusive, and they and other proposed mechanisms may, in fact, be acting to varying extents.

Indicators suggest that there has been an overall downturn in solar activity since the mid-1980s and that the Sun may currently be moving toward a state of lower activity. The occurrence of a future grand minimum, like the Maunder Minimum, within several decades is of low probability but cannot be ruled out. This raises the issue of whether the Sun might buy some time for the world to adjust to greenhouse gas–induced global warming. It would be rash, however, to become complacent on this basis for several reasons. First, predictions of solar activity are notoriously difficult and prone to error. Second, it is not necessarily the case that an Earth with a global net radiation balance but different radiative components (i.e., less absorbed solar radiation but more greenhouse trapping of infrared radiation) will have the same climate; indeed, some model studies of "geoengineering" approaches to mitigating climate change—which seek to artificially manage the amount of solar radiation entering the troposphere—have found significant differences in the hydrologic cycle and monsoons. Third, the time "bought" would probably be a decade or so at most, and on timescales of a few centuries the Sun is likely to return to a maximum state, resulting in a climate with more solar heating as well as a much higher greenhouse gas loading, and considerably warmer than at present.

The Sun—huge, beautiful, and constantly changing, is the source of life on Earth. The climate—complex, fascinating, and of fundamental importance to the human condition, depends on it. The relationship between the two provides scientific challenges that are both intrinsically engrossing and of great relevance to society, and we hope that in this book we have been able to convey at least some of the flavor of this subject.

Appendix   DETECTION OF SOLAR SIGNALS IN CLIMATE AND WEATHER RECORDS

THE DETECTION OF SIGNALS OF SOLAR ACTIVITY IN OB-servational records, such as described in Chapters 6 and 7, requires very careful analysis of the available data. The influence of the Sun needs to be separated from the effects of many other factors—including, for example, seasonal variations, volcanic eruptions, and ENSO—in data sets that inevitably incorporate observational errors, uncertainties, and gaps as well as inherent natural variability (noise). A wide variety of techniques have been used in this context; here we very briefly cover a few of the most common and direct the reader to some of the excellent texts available (see Suggestions for Further Reading) for more detailed information.

In all cases it is necessary to have some information about the temporal variability of solar activity and of any other proposed forcing factors. This information can be used in two fundamentally different ways. In the first approach the observational data are filtered, or otherwise analyzed, to produce signals on varying timescales, and the results are subsequently compared with the corresponding time series of solar activity, or some property of it.

A popular example of this approach is to identify periodicities in time series by performing a **harmonic analysis** (such as a Fourier transform) on the data. The result is a power spectrum showing the contributions of different frequency components (i.e., waves with different periods) to the time series. Spikes in the spectrum are then identified with particular forcing factors. Such analysis has been applied to many meteorological data sets and the results used, for example, to suggest 11-year sunspot cycles in U.S. rainfall and African lake levels (see Chapter 6). The problem with such analyses is that the signals are intermittent, so that, as far as we are aware, no robust solar signal has ever been demonstrated using this technique. More sophisticated extensions of harmonic analysis include the **wavelet transform**. This method, in which the infinitely long time series of single-frequency waves are replaced by small clusters of wavelets of defined lifetime, shows explicitly how features of a given spectral composition grow and decay in time.

Another example of this overall approach is **empirical mode decomposition**. In this technique smooth envelopes, defined by local maxima and minima in the time series, are constructed, and the mean of these two curves (called the first *intrinsic mode function* [IMD]) is subtracted from the data. The process is repeated, with each successive IMD representing an (irregular) oscillation of shorter temporal variability than the last until the residual time series is deemed to contain only noise. The IMDs are subsequently interpreted as associated with known forcing factors, such as solar activity, the quasi-biennial oscillation, or the El Niño–Southern Oscillation (an example is discussed in Chapter 7).

The advantage of techniques within this first approach is that for the filtering stage they require no preconceived ideas of specific influences, so that any signals fall naturally from the data analysis. The disadvantage is that the subsequent association of the derived signals with specific forcing factors is often based on correlations and subject to the same limitations (discussed next).

The second approach takes as input, alongside the observational data, time series of solar activity and other possible forcings. A simple example is just the **correlation** of a meteorological data set with a measure of solar activity, and the square of the correlation coefficient is taken as the fraction of the variability in the data set associated with the forcing factor. A simple **linear regression** is the logical next step, in which the magnitude of the variation in the data associated with that in the forcing index is deduced, and the statistical robustness of the relationship is determined from the number of data points, the magnitude of the signal, and the value of the correlation coefficient. This approach can be useful when one factor dominates the variability, and has been successfully applied to the extraction of solar signals in stratospheric temperatures (see Chapter 7).

An extension is **multiple linear regression** for situations in which it is assumed that several (known) influences are acting simultaneously. In its simplest form this method involves finding the weights of all forcing components such that their weighted sum is the best fit (given by the minimum sum of least-squares differences) to the times series being analyzed. Care has to be taken in determining the precision of the derived signals, including taking account of nonlinearity (such as introduced by correlations between the forcing indexes). This method has been used extensively to extract solar signals from observational data while accounting for other factors such as, for example, volcanic eruptions or stratospheric chlorine content; several examples are given in Chapters 6 and 7.

Methods under the general heading **optimal estimation** are used in the attribution of causes to climate change. These methods essentially involve least-squares fits to observational data with results from climate models used as the input functions. Importantly, they also take account of natural variability/noise in estimating the magnitude of the signal(s). Thus runs of a climate model are used to produce the responses to single forcing functions (e.g., greenhouse gases, solar irradiance) in time and space. A spatial pattern derived in this way is often referred to as the *fingerprint* of the forcing factor. Very long runs of the same model without any external forcings provide information on the structure of the noise. The component of each forcing is found by mapping the fingerprint, weighted by the noise, onto the observational data. Examples of the application of this method are presented in Chapters 6 and 7.

Results of statistical analyses must always be interpreted cautiously: whichever technique is used a solution is essentially worthless if it does not include a robust assessment of its limitations and uncertainty range.

Adiabatic—With no energy transfer between the system under consideration and its surroundings. A good approximation for air moving fairly rapidly, e.g., in convective cloud.

Aerosol—A suspension of airborne solid or liquid particles, with a typical size between a few nanometers and 10 μm, that resides in the atmosphere for at least several hours.

Albedo—The fraction of solar radiation reflected by a surface or object, often expressed as a percentage.

Anthropogenic—Resulting from human activity.

AOGCM—see Climate model

Atmosphere—The gaseous envelope surrounding the earth.

Auroras—Commonly visible as bright, fluctuating displays at high latitudes, auroras are the response of the ionosphere to energetic particles which, in turn, are accelerated during geomagnetic storms.

Carbon dioxide (CO_2)—A naturally occurring gas; also a by-product of burning fossil fuels, of land-use changes, and of industrial processes. It is the principal anthropogenic greenhouse gas.

Chlorofluorocarbons (CFCs)—Artificial organic compounds containing chlorine, carbon, hydrogen, and fluorine. CFCs have been used for refrigeration, air conditioning, packaging, plastic foam, insulation, solvents, and aerosol propellants.

Because they are not destroyed in the lower atmosphere, CFCs drift into the upper atmosphere, where, given suitable conditions, they break down ozone.

Climate—The average weather or, more rigorously, the statistical description in terms of the mean and variability of relevant quantities such as temperature, precipitation, and wind over a period of time ranging from months to thousands or millions of years.

Climate change—An alteration in the state of the climate that persists for an extended period, typically decades or longer. Climate change may result from natural forcings, such as modulations in solar radiation or volcanic eruptions, or from anthropogenic changes in the composition of the atmosphere or in land use.

Climate feedback—An interaction in which a perturbation in one climate quantity causes a change in a second, and the change in the second quantity ultimately leads to an additional change in the first. A negative feedback is one in which the initial perturbation is weakened by the changes it causes; a positive feedback is one in which the initial perturbation is enhanced.

Climate model—A numerical representation, of varying complexity, of the climate system, differing in such aspects as the number of spatial dimensions or the extent to which physical, chemical, or biological processes, and the interactions between them, are explicitly represented. A general circulation (global climate) model (GCM) includes a three-dimensional representation of the atmosphere and probably the oceans, along with components representing land and sea ice. Sometimes, acronyms make the components explicit, e.g., AGCM (atmosphere only), AOGCM (includes atmosphere–ocean coupling).

Climate sensitivity parameter—The equilibrium (steady-state) change in the annual global mean surface temperature in response to a unit change in radiative forcing (units: K $W^{-1}\,m^{-2}$).

Climate system—The highly complex system consisting of the atmosphere, the hydrosphere, the cryosphere, the lithosphere, and the biosphere, and the interactions among them.

Convection—Upward transport of air due to a statically unstable vertical temperature gradient, often resulting in the formation of cloud.

Coronal holes—Regions of open magnetic field in the solar corona. They are the origin of the faster components of the solar wind.

Coronal mass ejection (CME)—An eruption of plasma and magnetic flux from the Sun. Its interaction with the Earth's magnetosphere can lead to major geomagnetic storms.

Cosmogenic nuclide—An isotope produced by the collision of high-energy cosmic rays with the nucleus of an atom. The incidence of cosmic rays on the atmosphere is modulated by the solar magnetic field, so that records of, e.g., ^{10}Be in ocean and ice sediments can be used as a measure of solar activity.

Eccentricity—A measure of the deviation of the shape of the Earth's orbit around the Sun from circular to elliptical.

ENSO (El Niño–Southern Oscillation)—A natural variation in the atmosphere and ocean of the tropical Pacific indicated by warm/cold sea surface temperatures and low/high sea-level pressure in the eastern/western Pacific (the positive phase; El Niño), and the reverse in the negative phase (La Niña).

F10.7—Radiation of 10.7 cm (radio) wavelength due to a combination of plasma processes that are more important for strong magnetic fields and high densities; a proxy for solar UV radiation that correlates positively with other activity.

Faculae—Small, bright regions seen on the Sun's surface. They are associated with regions of strong magnetic field and are a net positive contribution to TSI at the maximum of the solar cycle.

Fossil fuels—Carbon-based fuels from fossil hydrocarbon deposits, including coal, peat, oil, and natural gas.

Galactic cosmic rays (GCRs)—Energetic charged particles (hundreds of mega-electron volts per nucleon). They are believed to be accelerated within the galaxy. Their flux at the Earth is inversely correlated with other manifestations of the solar cycle (i.e., maximum flux at minimum of the cycle). On entering the Earth's atmosphere they collide with molecules, producing a cascade of lighter particles including muons, which can be measured at the surface to indicate the intensity of incident GCRs.

GCM—See Climate model

Geomagnetic field—The magnetic field of the Earth. The field strength at the surface is of order 5×10^{-5} T.

Geomagnetic storm—A disturbance of the geomagnetic field due to solar activity. The largest storms are due to coronal mass ejections originating at the Sun. Typical field changes in large storms are of order 200 nT. During a storm, the magnetosphere becomes compressed, the ring currents are enhanced, and auroras are common.

Geopotential height—The altitude of a specified pressure level. It indicates the mean temperature of the air below that level.

Greenhouse effect—The heating effect exerted by the atmosphere on the Earth because the atmosphere, being relatively transparent to solar radiation, allows it to reach and warm the surface but being absorptive in the infrared, absorbs heat (infrared radiation) emitted by the Earth's surface and traps it at low levels.

Greenhouse gases (GHGs)—Gaseous constituents of the atmosphere, both natural and anthropogenic, that absorb and emit radiation at wavelengths within the spectrum of infrared radiation emitted by the Earth's surface and by the atmosphere itself.

Hadley cell—An atmospheric overturning circulation between the tropics and subtropics with poleward flow near the tropopause and equatorward flow near the surface.

Helioseismology—A technique for studying the structure of the solar interior based on the analysis of small surface oscillations that are the manifestation of sound waves.

Ice core—A long cylindrical sample taken through the depth of an ice sheet. Annual layers of snowfall leave a record of atmospheric temperature and composition that can be identified from a core as far back as 800,000 years.

Insolation—Total amount of solar radiation energy received on a given surface area during a given time

Intertropical convergence zone (ITCZ)—A region around the Earth in the tropics where the northeast and southeast trade winds meet, causing the air to rise and resulting in clouds and thunderstorms. It moves north and south seasonally, following the Sun.

Ionization—The process whereby one or more electrons are removed from an atom or molecule.

Ionosphere—A weakly ionized region roughly 100–300 km above the surface of the Earth.

Irradiance—Radiant energy per unit time crossing unit area perpendicular to its direction of travel (units: W m^{-2}).

Jet stream—A strong meandering current of air flowing in an overall west-to-east direction near the tropopause in midlatitudes.

Lower atmosphere—The troposphere and lower stratosphere.

Magnetosphere—A magnetic cavity in space carved out of the solar wind by the Earth's magnetic field. It extends roughly 10 R$_E$ in the sunward direction and up to 100 R$_E$ on the night side. The plasma density in the magnetosphere is small compared with that in the solar wind.

Meridional—The north-south direction.

Mesosphere—The layer of the atmosphere between about 50 and 80 km altitude in which temperature decreases with height.

Middle atmosphere—The upper stratosphere and mesosphere.

Monsoon—A change in atmospheric circulation and precipitation in the tropics associated with seasonal variations in the land–sea temperature contrast. The intense monsoon rainfall is responsible for a large fraction of the annual precipitation in many regions affected, and the date of onset is important for agriculture.

North Atlantic Oscillation (NAO)—A natural variation in the meridional surface pressure gradient over the North Atlantic Ocean. In its positive phase the NAO is associated with a stronger, more zonal, and more northerly jet stream that brings milder, wetter weather to northwestern Europe.

Northern Annular Mode (NAM)—A hemispheric extension of the NAO in which the strong westerly winds blowing around northern mid–high latitudes throughout the troposphere and stratosphere move toward (positive phase) or away from (negative phase) the pole.

Obliquity—The tilt of the Earth's axis relative to the perpendicular to the plane of its orbit around the Sun.

Open magnetic flux—The magnetic flux (formally) measured in the solar wind that originates at the Sun. It shows positive correlation with other manifestations of solar activity.

Photodissociation—The fragmentation of a molecule into its components after absorption of the energy of an incident photon.

Plasma—An ionized gas in which electromagnetic forces dominate the behavior.

Precession—The change in orientation of the Earth's axis. It can be measured as the position on the Earth's elliptical orbit of the spring equinox.

Radiation (electromagnetic)—Transfer of energy through perpendicular oscillating electric and magnetic fields. The spectrum extends from very long wavelengths ($\sim 10^4$ m) through radio waves, microwaves, infrared radiation, visible light, ultraviolet radiation, X-rays, and gamma rays to extremely short wavelengths ($\sim 10^{-16}$ m). It can also be viewed as consisting of particles (photons) ranging from low to high energy.

Radiation (ionizing)—Particles, including cosmic rays, that have sufficient energy to liberate electrons from atoms or molecules.

Radiative forcing—The change in the net (downward minus upward) radiative flux (expressed in W m^{-2}) at the top of the atmosphere, or at the tropopause, that results from an instantaneous change in a driver such as the concentration of carbon dioxide or the output of the Sun.

Reanalysis—A data set in which a climate model has been used with observational data to fill gaps and improve internal consistency.

Saturation vapor pressure—The vapor pressure at which air can hold no more water vapor. The value of the saturated vapor pressure increases with the temperature of the air.

Solar activity—Indications that the Sun is not a constant body, including the formation and decay of sunspots, solar flares, coronal mass ejections, and the emission of particles as well as variations in emission of electromagnetic radiation.

Solar chromosphere—A thin layer between the photosphere and corona with temperatures between ∼ 6000 and 30,000 K. The chromosphere must be heated in situ by a range of plasma processes that dissipate kinetic and/or magnetic energy.

Solar corona—The outer solar atmosphere. Its temperature is greater than 1 MK, and it comprises magnetically open (solar wind) and closed regions. Radiation from the latter is dominated by active regions at the peak of the solar cycle.

Solar cycle—The cycle of the solar magnetic field. The magnetic field reverses polarity roughly every 11 years, so a complete cycle takes of order 22 years. More commonly, the term refers to 11 years. Sunspot number, F10.7 flux, and solar and geomagnetic activity all exhibit an 11-year cycle.

Solar photosphere—The visible surface of the Sun.

Solar wind—The expansion of the solar corona into interplanetary space. At the Earth it is highly supersonic (hundreds of kilometers per second). The Sun's open magnetic field permeates the wind with typical strength at the Earth of a few nanotesla.

Southern Annular Mode (SAM)—A naturally occurring variation in which the strong westerly winds blowing around southern mid–high latitudes throughout the troposphere and stratosphere move toward (positive phase) or away from (negative phase) the pole.

Space weather—A general term for disturbed electromagnetic and plasma conditions, and enhanced energetic particle populations in the magnetosphere, ionosphere, and terrestrial upper atmosphere and surface due to solar activity or GCRs.

Storm tracks—Preferred routes for midlatitude cyclones, which tend to travel west to east across the oceans in both hemispheres.

Stratosphere—The region of the atmosphere above the troposphere extending from about 10 km (lower near the poles and higher in the tropics) to about 50 km altitude. Temperatures generally increase with height, making it very stable.

Sunspot—A region of intense (>0.1 T) magnetic field visible on the solar surface. Sunspots are cooler than the visible surface and are a net negative contribution to TSI at the maximum of the solar cycle. They have a lifetime of days or weeks; the latitudinal position and number of sunspots varies with the solar cycle.

Sunspot number—In simple terms, an estimate of the level of solar activity as measured by both the number of sunspot groups and individual spots. Alternative measurements consider the area of the spots.

Tree ring—The annual layer of wood produced by a tree trunk as determined by local temperature and humidity. A cross section shows rings of varying thickness, which can be analyzed as a climate record. Although precise dating can be a challenge, wood dating back more than 10,000 years has been used to indicate past climate.

Tropopause—The boundary between the troposphere and stratosphere.

Troposphere—The lowest part of the atmosphere, from the surface to about 10 km in altitude (lower near the poles and higher in the tropics), where clouds and weather phenomena occur. In the troposphere, temperatures generally decrease with height.

TSI (Total solar irradiance)—Solar radiant power integrated over all wavelengths, per unit area perpendicular to the beam, at the Earth's distance from the Sun. Current value is 1361 W m^{-2}.

Uncertainty—Incomplete knowledge that can result from, e.g., imprecise or gaps in data, ambiguous definitions, or intrinsic variability. Uncertainty can be represented by quantitative measures (e.g., a probability density function or range of results) or by qualitative statements.

Upper atmosphere—The region above the mesopause.

Vapor pressure—A measure of humidity, the partial pressure exerted by the water vapor in the air.

Walker cell—A large-scale overturning of the atmosphere in the tropical Pacific with westward flow along the surface and eastward flow aloft. It is strongly associated with ENSO and with variations in sea surface temperatures as well as with the intensity and location of rainfall over Indonesia.

Weather—A description of the meteorological state (temperature, pressure, wind, humidity, precipitation, cloudiness, sunshine) at a particular place and time.

Zonal—The east-west direction.

Suggestions for Further Reading

··

Atmosphere and Climate Fundamentals (Chapters 2 and 4)

J. P. Peixoto and A. H. Oort. *The Physics of Climate*. Springer, 1997.
> Covers key features of the climate system, with physical and mathematical background suitable for students and professionals.

G. W. Petty. *A First Course in Atmospheric Radiation*. Sundog, 2011.
> Comprehensive coverage of atmospheric radiative transfer at the undergraduate/graduate level.

D. A. Randall. *Atmosphere, Clouds, and Climate*. Princeton University Press, 2012.
> Readable introduction to the climate system for the nonspecialist.

D. A. Randall (Ed.). *General Circulation Model Development: Past, Present, and Future*. Academic Press, 2000.
> Each chapter describes a different aspect of the history, design, development, or component of GCMs. Provides detailed and mathematical coverage.

T. F. Stocker, D. Qin, G.-K. Plattner, M. Tignor, K. Allen, J. Boschung, A. Nauels, Y. Xia, V. Bex, and P.M. Midgley (Eds.). Climate Change 2013: The physical science basis. Contribution of Working Group I to the Fifth Assessment

··

Report of the Intergovernmental Panel on Climate Change. CUP, 2013. http://www.ipcc.ch/report/ar5/wg1/.

> The IPCC reports present an overview of the current state of understanding of climate science (Working Group I), climate change impacts (WG II), and mitigation pathways (WG III). The full reports are very detailed; each contains an Executive Summary, Technical Summary, and Summary for Policymakers aimed at different readerships.

J. M. Wallace and P. V. Hobbs. *Atmospheric Science: An Introductory Survey*, 2nd ed. Academic Press, 2006.

> Covers physical processes in the atmosphere and climate at undergraduate level.

N. C. Wells. *The Atmosphere and Ocean: A Physical Introduction*, 3rd. ed. Wiley-Blackwell, 2011.

> Focuses on the interdependence of the atmosphere and oceans.

The Sun and Space Weather (Chapters 3, 5, and 8)

Annual Reviews of Astronomy and Astrophysics (http://www.annualreviews.org/loi/astro).

> Contains occasional reviews of interest. Gizon et al. (2010) cover helioseismology, Zhang and Low (2005) cover coronal mass ejections, and Thomas and Weiss (2004) cover sunspots.

M. Kivelson and C. Russell. *Introduction to Space Physics*. Cambridge University Press, 1995.

> Introduction to the Sun, solar wind, and magnetospheres.

Y. Kamide and A. Chian (Eds.). *Handbook of the Solar Terrestrial Environment*. Springer, 2007.

> Provides an overview of solar and solar terrestrial physics, with chapters on space weather.

Living Reviews in Solar Physics.

An online journal that publishes a few comprehensive review articles each year. In recent years there have been many articles of direct relevance to this book.

E. R. Priest. *Magnetohydrodynamics of the Sun.* Cambridge University Press, 2014.

A comprehensive introduction to the physics of the outer solar atmosphere, largely from a theoretical perspective.

UK Royal Academy of Engineering. Extreme space weather: Impacts on engineered systems and infrastructure. 2013. http://www.raeng.org.uk/societygov/policy/current_issues /space_weather/default.htm.

A UK advocacy document for a program to study, forecast, and mitigate large space weather events.

F. Shu. *The Physics of Astrophysics.* Volume 2: *Gas Dynamics.* University Science Books, 1992.

A useful introduction to stellar structure.

Solar Signals in Climate (Chapters 6 and 7)

W. J. Burroughs. *Weather Cycles: Real or Imaginary*? Cambridge University Press, 2003.

Readable discussion of the history and practice of the search for solar signals in and weather and climate records.

L. J. Gray et al. Solar influence on climate. *Reviews of Geophysics* **48**, RG4001, doi:10.1029/2009RG000282, 2010.

Overview of scientific state-of-the art (ca. 2009) in solar–climate links.

D. V. Hoyt and K. H. Schatten *The Role of the Sun in Climate Change.* Oxford University Press, 1997.

Readable and knowledgeable perspective on variations in the Sun the search for solar links in climate.

J. M. Pap and P. Fox (Eds.). Solar variability and its effects on the Earth's atmospheric and climate system. *Geophysical Monograph* **141**, 2004.
Each chapter presents an overview of the science in a different aspect of solar and climate variability.

S. M. Solanki et al. Solar irradiance variability and climate. *Annual Review of Astronomy and Astrophysics* **51**, 311–351, 2013.
Technical review article.

Statistical Techniques (Appendix)

C. Duchon and R. Hale. *Time Series Analysis in Meteorology and Climatology: An Introduction*. Wiley-Blackwell, 2012.
Pedagogical text on spectral analysis with mathematical background and examples of applications.

H. Kaper and H. Engler. *Mathematics and Climate*. Society for Industrial and Applied Mathematics, 2013.
Designed around a wide range of climate themes, the math and statistics are presented in a user-friendly manner.

H. von Storch and F. W. Zwiers. *Statistical Analysis in Climate Research*. Cambridge University Press, 2002.
A comprehensive coverage of relevant statistical approaches.

Bibliography

1 Cook, K. H. *Climate Dynamics*. (Princeton University Press, 2013).

2 Salby, M. L. *Fundamentals of Atmospheric Physics*. (Academic Press, 1996).

3 Houghton, J. T. *The Physics of Atmospheres*, 3rd ed. (Cambridge University Press, 2002).

4 Laing, A., & Evans, J.-L. *Introduction to Tropical Meteorology*. (UCAR, 2011).

5 Intergovernmental Panel on Climate Change (IPCC). Climate change 2007: The physical science basis. Contribution of Working Group I to the Fourth Assessment Report of the Intergovernmental Panel on Climate Change. (2007).

6 Lean, J. L., & Rind, D. Climate forcing by changing solar radiation. *Journal of Climate* **11**, 3069–3094 (1998).

7 Shu, F. *The Physical Universe*. (University Science Books, 1985).

8 Gizon, L., Birch, A. C., & Spruit, H. C. Local helioseismology: Three-dimensional imaging of the solar interior. *Annual Review of Astronomy and Astrophysics* **48**, 289–338 (2010).

9 Parker, E. N. Hydromagnetic dynamo models. *Astrophysical Journal* **121**, 293 (1955).

10 Thornton, L. M., & Parnell, C. E. Small-scale flux emergence observed using *Hinode*/SOT. *Solar Physics* **269**, 13–40, doi:10.1007/s11207–010–9656–7 (2011).

11 Thomas, J. H., & Weiss, N. O. Fine structure in sunspots. *Annual Review of Astronomy and Astrophysics* **42**, 517–548, doi:10.1146/annurev.astro.42.010803.115226 (2004).

12 Foukal, P., Frohlich, C., Spruit, H., & Wigley, T.M.L. Variations in solar luminosity and their effect on the Earth's climate. *Nature* **443**, 161–166, doi:10.1038/nature05072 (2006).

13 Klimchuk, J. A., Patsourakos, S., & Cargill, P. J. Highly efficient modeling of dynamic coronal loops. *Astrophysical Journal* **682**, 1351–1362, doi:10.1086/589426 (2008).

14 Cargill, P., & De Moortel, I. Solar physics: Waves galore. *Nature* **475**, 463–464 (2011).

15 Withbroe, G. L., & Noyes, R. W. Mass and energy flow in solar chromosphere and corona. *Annual Review of Astronomy and Astrophysics* **15**, 363–387, doi:10.1146/annurev.aa.15.090177.002051 (1977).

16 Parker, E. N. Dynamics of the interplanetary gas and magnetic fields. *Astrophysical Journal* **128**, 664 (1958).

17 Dennis, B. R., Emslie, A. G., & Hudson, H. S. Overview of the volume. *Space Science Reviews* **159**, 3–17, doi:10.1007/s11214–011–9802-z (2011).

18 Zhang, M., & Low, B. C. The hydromagnetic nature of solar coronal mass ejections. *Annual Review of Astronomy and Astrophysics* **43**, 103–137 (2005).

19 Berger, A., & Loutre, M. F. Insolation values for the climate of the last 10 million years. *Quaternary Science Reviews* **10**, 297–317, doi:10.1016/0277–3791(91)90033-q (1991).

20 Randall, D. A. *Atmosphere, Clouds, and Climate.* (Princeton University Press, 2012).

21 IPCC. Climate change 2013: The physical science basis. Contribution of Working Group I to the Fifth Assessment Report of the Intergovernmental Panel on Climate Change, 1535 (2013).

22 Goody, R. M., & Yung, Y. L. *Atmospheric Radiation* (Oxford University Press, 1989).

23 Gray, L. J., et al. Solar influences on climate. *Reviews of Geophysics* **48**, RG4001, (2010).

24 Fröhlich, C., & Lean, J. The Sun's total irradiance: Cycles, trends and related climate change uncertainties since 1976. *Geophysical Research Letters* **25**, 4377–4380 (1998).

25 Tapping, K. F. Recent solar radio astronomy at centimeter wavelengths: The temporal variability of the 10.7-cm flux. *Journal of Geophysical Research: Atmospheres* **92**, 829–838, doi:10.1029/JD092iD01p00829 (1987).

26 Henney, C. J., Toussaint, W. A., White, S. M., & Arge, C. N. Forecasting F-10.7 with solar magnetic flux transport modeling. *Space Weather: The International Journal of Research and Applications* **10**, doi:10.1029/2011sw000748 (2012).

27 McComas, D. J., et al. The three-dimensional solar wind around solar maximum. *Geophysical Research Letters* **30**, doi:10.1029/2003gl017136 (2003).

28 Lockwood, M., & Stamper, R. Long-term drift of the coronal source magnetic flux and the total solar irradiance. *Geophysical Research Letters* **26**, 2461–2464, doi:10.1029/1999gl900485 (1999).

29 Gray, L., et al. Solar influences on climate. See reference 23.

30 Solanki, S. K., & Fligge, M. Reconstruction of past solar irradiance. *Space Science Reviews* **94**, 127–138, doi:10.1023/a:1026754803423 (2000).

31 Lean, J., Beer, J. & Bradley, R. Reconstruction of solar irradiance since 1610: Implications for climate change. *Geophysical Research Letters* **22**, 3195–3198 (1995).

32 Krivova, N. A., Vieira, L E.A., & Solanki, S. K. Reconstruction of solar spectral irradiance since the Maunder minimum. *Journal of Geophysical Research: Space Physics* **115**, doi:10.1029/2010ja015431 (2010).

33 Steinhilber, F., Beer, J., & Frohlich, C. Total solar irradiance during the Holocene. *Geophysical Research Letters* **36**, doi:10.1029/2009gl040142 (2009).

34 Shapiro, A. I., et al. A new approach to the long-term reconstruction of the solar irradiance leads to large historical solar forcing. *Astronomy & Astrophysics* **529**, doi:10.1051/0004 -6361/201016173 (2011).

35 Lockwood, M., Owens, M., Barnard, L., Davis, C., & Thomas, S. Solar cycle 24: What is the Sun up to? *Astronomy & Geophysics* **53**, 9–15, doi:10.1111/j.1468-4004.2012.53309.x (2012).

36 Stauffer, B. Long term climate records from polar ice. *Space Science Reviews* **94**, 321–336, doi:10.1023/a:1026791811601 (2000).

37 Bond, G. *et al.* Persistent solar influence on north Atlantic climate during the Holocene. *Science* **294**, 2130–2136 (2001).

38 IPCC. Climate change 2007. See reference 5.

39 Lean, J. L., & Rind, D. H. How natural and anthropogenic influences alter global and regional surface temperatures: 1889 to 2006. *Geophysical Research Letters* **35**, doi:L18701 10.1029 /2008gl034864 (2008).

40 Schurer, A. P., Hegerl, G. C., Mann, M. E., Tett, S. F. B. & Phipps, S. J. Separating forced from chaotic climate variability over the past millennium. *Journal of Climate* **26**, 6954–6973, doi:10.1175/jcli-d-12-00826.1 (2013).

41 IPCC. Climate change 2013. See reference 21.

42 Meehl, G. A., Arblaster, J. M., & Marsh, D. R. Could a future "Grand Solar Minimum" like the Maunder Minimum stop global warming? *Geophysical Research Letters* **40**, 1789–1793, doi:10.1002/grl.50361 (2013).

43 Currie, R. G. Luni-solar 18.6-year and solar-cycle 10–11-year signals in USA air temperature records. *International Journal of Climatology* **13**, 31–50, doi:10.1002/joc.3370130103 (1993).

44 Brown, G. M., & John, J. I. Solar-cycle influences in tropospheric circulation. *Journal of Atmospheric and Terrestrial Physics* **41**, 43–52, doi:10.1016/0021-9169 (1979).

45 Cohen, T. J., & Sweetser, E. I. The "spectra" of the solar cycle and of data for Atlantic tropical cyclones. *Nature* **256**, 295–296, doi:10.1038/256295a0 (1975).

46 Hoyt, D. V., & Schatten, K. H. *The Role of the Sun in Climate Change*. (Oxford University Press, 1997).

47 Stevens, M. J., & North, G. R. Detection of the climate response to the solar cycle. *Journal of the Atmospheric Sciences* **53**, 2594–2608, doi:10.1175/1520-0469 (1996).

48 Mann, M. E., et al. Global signatures and dynamical origins of the Little Ice Age and Medieval climate anomaly. *Science* **326**, 1256–1260, doi:10.1126/science.1177303 (2009).

49 Woollings, T., Lockwood, M., Masato, G., Bell, C., & Gray, L. Enhanced signature of solar variability in Eurasian winter climate. *Geophysical Research Letters* **37**, doi:10.1029/2010gl044601 (2010).

50 Gray, L. J., et al. A lagged response to the 11 year solar cycle in observed winter Atlantic/European weather patterns. *Journal of Geophysical Research: Atmospheres* **118**, 13405–13420, doi:10.1002/2013jd020062 (2013).

51 Christoforou, P., & Hameed, S. Solar cycle and the Pacific "centers of action." *Geophysical Research Letters* **24**, 293–296 (1997).

52 Roy, I., & Haigh, J. D. Solar cycle signals in sea level pressure and sea surface temperature. *Atmospheric Chemistry and Physics* **10**, 3147–3153 (2010).

53 Roscoe, H. K., & Haigh, J. D. Influences of ozone depletion, the solar cycle and the QBO on the Southern Annular Mode. *Quarterly Journal of the Royal Meteorological Society* **133**, 1855–1864, doi:10.1002/qj.153 (2007).

54 Haigh, J. D. The effects of solar variability on the Earth's climate. *Philosophical Transactions of the Royal Society of London Series A: Mathematical, Physical & Engineering Sciences* **361**, 95–111 (2003).

55 van Loon, H., Meehl, G. A., & Shea, D. J. Coupled air-sea response to solar forcing in the Pacific region during northern winter. *Journal of Geophysical Research: Atmospheres* **112**, doi:D02108 10.1029/2006jd007378 (2007).

56 van Loon, H., & Meehl, G. A. The Indian summer monsoon during peaks in the 11 year sunspot cycle. *Geophysical Research Letters* **39**, doi:10.1029/2012gl051977 (2012).

57 Meehl, G. A., Arblaster, J. M., Branstator, G., & van Loon, H. A coupled air-sea response mechanism to solar forcing in the Pacific region. *Journal of Climate* **21**, 2883–2897, doi:10.1175/2007jcli1776.1 (2008).

58 Misios, S., & Schmidt, H. Mechanisms involved in the amplification of the 11-yr solar cycle signal in the tropical Pacific Ocean. *Journal of Climate* **25**, 5102–5118, doi:10.1175/jcli-d-11-00261.1 (2012).

59 Emmert, J. T., Lean, J. L., & Picone, J. M. Record-low thermospheric density during the 2008 solar minimum. *Geophysical Research Letters* **37**, doi:10.1029/2010gl043671 (2010).

60 Labitzke, K., & Vanloon, H. Associations between the 11-year solar-cycle, the quasi-biennial oscillation and the atmosphere: A summary of recent work. *Philosophical Transactions of the Royal Society of London Series A: Mathematical, Physical & Engineering Sciences* **330**, 577–589 (1990).

61 Labitzke, K. On the solar cycle–QBO relationship: A summary. *Journal of Atmospheric and Solar-Terrestrial Physics* **67**, 45–54 (2005).

62 Randel, W. J., et al. An update of observed stratospheric temperature trends. *Journal of Geophysical Research: Atmospheres* **114**, doi:10.1029/2008jd010421 (2009).

63 Chiodo, G., Marsh, D. R., Garcia-Herrera, R., Calvo, N., & Garcia, J. A. On the detection of the solar signal in the tropical stratosphere. *Atmospheric Chemistry and Physics* **14**, 5251–5269, doi:10.5194/acp-14-5251-2014 (2014).

64 Haigh, J. D. (2003). See reference 54.

65 Coughlin, K., & Tung, K. K. Eleven-year solar cycle signal throughout the lower atmosphere. *Journal of Geophysical Research: Atmospheres* **109**, doi:10.1029/2004jd004873 (2004).

66 Frame, T.H.A., & Gray, L. J. The 11-yr solar cycle in ERA-40 data: An update to 2008. *Journal of Climate* **23**, 2213–2222, doi:10.1175/2009jcli3150.1 (2010).

67 Haigh, J. D., & Blackburn, M. Solar influences on dynamical coupling between the stratosphere and troposphere. *Space Science Reviews* **125**, 331–344 (2006).

68 IPCC. Climate change 2001: The scientific basis. Contribution of Working Group I to the Third Assessment Report of the Intergovernmental Panel on Climate Change, 881 (2001).

69 Randel, W. J., & Wu, F. A stratospheric ozone profile data set for 1979–2005: Variability, trends, and comparisons with column ozone data. *Journal of Geophysical Research: Atmospheres* 112, doi:D06313 (2007).

70 Hood, L. L., & Soukharev, B. E. The lower-stratospheric response to 11-yr solar forcing: Coupling to the troposphere-ocean response. *Journal of the Atmospheric Sciences* 69, 1841–1864, doi:10.1175/jas-d-11-086.1 (2012).

71 Dikty, S., et al. Modulations of the 27 day solar rotation signal in stratospheric ozone from Scanning Imaging Absorption Spectrometer for Atmospheric Cartography (SCIAMACHY) (2003–2008). *Journal of Geophysical Research: Atmospheres* 115, doi:10.1029/2009jd012379 (2010).

72 Forster, P. M., et al. Evaluation of radiation scheme performance within chemistry climate models. *Journal of Geophysical Research: Atmospheres* 116, doi:10.1029/2010jd015361 (2011).

73 Harder, J. W., Fontenla, J. M., Pilewskie, P., Richard, E. C., & Woods, T. N. Trends in solar spectral irradiance variability in the visible and infrared. *Geophysical Research Letters* 36, doi:L07801 10.1029/2008gl036797 (2009).

74 Haigh, J. D., Winning, A. R., Toumi, R., & Harder, J. W. An influence of solar spectral variations on radiative forcing of climate. *Nature* 467, 696–699, doi:10.1038/nature09426 (2010).

75 Rusch, D. W., Gerard, J. C., Solomon, S., Crutzen, P. J., & Reid, G. C. The effect of particle precipitation events on the neutral and ion chemistry of the middle atmosphere. I. Odd nitrogen. *Planetary and Space Science* 29, 767–774, doi:10.1016/0032-0633(81)90048-9 (1981).

76 Solomon, S., Rusch, D. W., Gerard, J. C., Reid, G. C., & Crutzen, P. J. The effect of particle-precipitation events on the neutral and ion chemistry of the middle atmosphere. 2. Odd hydrogen. *Planetary and Space Science* **29**, 885–892, doi:10.1016/0032-0633(81)90078-7 (1981).

77 Jackman, C. H., et al. Long-term middle atmospheric influence of very large solar proton events. *Journal of Geophysical Research: Atmospheres* **114**, doi:10.1029/2008jd011415 (2009).

78 Harrison, R. G., & Carslaw, K. S. Ion-aerosol-cloud processes in the lower atmosphere. *Reviews of Geophysics* **41**, doi:1012 10.1029/2002rg000114 (2003).

79 Duplissy, J., et al. Results from the CERN pilot CLOUD experiment. *Atmospheric Chemistry and Physics* **10**, 1635–1647 (2010).

80 Ineson, S., et al. Solar forcing of winter climate variability in the Northern Hemisphere. *Nature Geoscience* **4**, 753–757, doi:10.1038/ngeo1282 (2011).

81 Matthes, K., Kuroda, Y., Kodera, K., & Langematz, U. Transfer of the solar signal from the stratosphere to the troposphere: Northern winter. *Journal of Geophysical Research: Atmospheres* **111**, doi:10.1029/2005jd006283 (2006).

82 Haigh, J. D. The impact of solar variability on climate. *Science* **272**, 981–984 (1996).

83 Ineson, S., et al. (2011). See reference 80.

84 Haynes, P. Stratospheric dynamics. *Annual Review of Fluid Mechanics* **37**, 263–293 (2005).

85 Gray, L. J., et al. (2010). See reference 23.

86 Meehl, G. A., Arblaster, J. M., Matthes, K., Sassi, F. & van Loon, H. Amplifying the Pacific climate system response to a

small 11-year solar cycle forcing. *Science* **325**, 1114–1118, doi:10 .1126/science.1172872 (2009).

87 Dungey, J. W. Interplanetary magnetic field and the auroral zones. *Physical Review Letters* **6**, 47 (1961).

88 Ibid.

89 Kamide, Y. & Chian, A. (Eds.). *Handbook of the Solar Terrestrial Environment*. (Springer, 2007).

90 Cannon, P. Extreme space weather: Impacts on engineered systems and infrastructure. (UK Royal Academy of Engineering, http://www.raeng.org.uk/societygov/policy/current _issues/space_weather/default.htm, 2013).

Index

floods, 5
forest fires, 5

galactic cosmic rays (GCRs), 17,
103–4, 106, 107, 108, 161;
galactic cosmic ray (GCR) flux,
101, 165; as a major nonsolar
space weather hazard, 156; role
of in atmospheric ionization,
140, 142–43, 152–53; single
cosmic rays and the scrambling
of computer memory, 162
Galileo, 2
Gauss, Carl Friedrich, 4
general circulation models. *See*
global climate models (HCMs)
"geoengineering," 169
geomagnetism, 7; geomagnetic
indexes, 8; geomagnetic storms,
4, 158, 159; geomagnetically
induced currents, 160–61
geopotential height, 19
global climate models (GCMs),
31, 34, 35, 76, 120, 128;
atmosphere-ocean GCMs, 125;
constraints on the performance
of, 35–36; and models of
reduced complexity, 38–39;
radiative processes in, 36–37;
response of storm tracks in to
changes in UV radiation, 146;
water and clouds in, 37–38
global electric circuit, the, and near-
cloud aerosol ionization, 152–53
global warming, 9–10, 117; "hiatus"
in, 122–23
great solar minimum (2008/2010),
107–8, 165
greenhouse effect, 72–74, 84
greenhouse gases, 24, 31, 73, 74,
76, 84, 117, 120, 167, 174;

anthropogenic greenhouse
gases, 10, 113; radiative forcing
from, 122
Greenland, 110, 124; glaciers of, 114

Hadley cell/circulation, 22, 24, 126,
152
heat, transfer of by circulation of the
atmosphere, 22–23

helioseismology, 48–49, 51
Herschel, William, 2, 4, 7
Holocene: isotope records of, 114;
proxy temperature records of,
124–25
humidity, 16, 38, 109–10, 127
hydrologic cycle, the, 37, 169

ice ages, 112; mechanism for,
113–14
ice "anvils," 16
ice cores, 9, 104, 110, 114
ice sheets: of the Northern Hemisphere, 112; thinning of Arctic
ice sheets, 116
icebergs, in the North Atlantic, 114
India/Indian subcontinent, famines
in, 6
infrared radiation (IR), 14, 22, 45,
82, 83, 84; infrared radiative
cooling, 18, 85; radiation in solar
IR lines, 60
insolation, 1, 77, 113, 138
interplanetary magnetic field (IMF),
65, 100–103, 104, 156; the
southward IMF and magnetic
reconnection, 158
intertropical convergence zone
(ITCZ), 16, 21, 22, 24, 127
intrinsic mode function (IMD), 172
ion-ion recombination, 142